U0322271

World Landscape Case Studies

THE BEST VILLA GARDEN LANDSCAPE

Edited by HI-DESIGN PUBLISHING

Dalian University of Technology Press

World Landscape Case Studies
THE BEST VILLA GARDEN LANDSCAPE

Published by
Dalian University of Technology Press

Address: Section B, Sic-Tech Building, No.80 Software
Park Road, Ganjingzi District, Dalian, China
Tel: +86 411 84709043
Fax: +86 411 84709246
E-mail: yuanbinbooks@gmail.com
URL: http://www.dutp.cn

ISBN 978-7-5611- 9151-4

PREFACE 1

Scott Brown
Scott Brown Landscape Design

Landscape Design – or Garden Design – can mean many different things to different people. Essentially it can represent a people's culture at a particular stage in their evolution, in much the same way as architecture does. Gardens are as much an expression of a people's influences, lifestyle, priorities, aspirations and the conditions under which they live as their architecture, and their cuisine. And just as in architecture and native cuisine, there have been changes in influence and style throughout human existence and this trend continues today.

In South-Eastern Australia, there has been a drought which broke two years ago. This event had a significant effect on people's attitude to gardens in general. In fact sustainability and water conservation in general has become a far more main stream focus in people's minds.

Perhaps the most significant trend within Australia's capital cities over the last 6-7 years is the increased profile of Outdoor Living and Al Fresco Dining. With increased gentrification in various parts of Melbourne, and smaller outdoor spaces being more prevalent with inner city suburban living, the importance of extending the indoors-outwards to maximise our enjoyment of our homes has emerged. This popularity of outdoor 'kitchens', and entertaining areas has spread to all areas of Melbourne – even the outlying suburbs now have new home package designs which include the 'Al Fresco dinning courtyard experience'. So this Al Fresco phenomenon has now become fashionable as well as being born of necessity (due to a lack of space and a need to minimise water usage).

Overall, I have seen the profile of Landscape/garden design increase dramatically over the last 15 years (especially the last 7). As people have become more aware of their surroundings and have been exposed to what is possible, the standard of design has increased, and this process continues. Essential elements in the garden (which might be a vegetable garden for one family) – which would have simply been 'plonked' in the sunniest part of the backyard – are now being designed so that they are attractive and connect with the rest of the outdoor environment. Our increasing consciousness is being mirrored by the continual evolution of our gardens.

PREFACE 2

Merilen Mentaal
MentaalLandscapes

In an increasingly globalizing world like ours, where almost every aspect of life is experiencing rapid changes, technology development, cross-national influence and widening social networking, gardens are influenced by that same wind of changes. They are mixing together the elements of ancient wisdom, proved truths and trialling new possibilities.

This book gives you an excellent opportunity to discover what the world of garden design is experiencing, where it is heading, and how similar or different the gardens are made in different countries.

Look at the concepts of the gardens, the houses that determine their style, the hidden but necessary framework, inspect the materials that have been used, and discover the planting that always has so much to say about a garden and its owners. The planting styles are very different, from strictly controlled solitary plants to a wild mixture of meadow planting with innumerable possibilities in between, creating masterpieces from trees, shrubs, perennial and annual plants. Find the partly hidden elements, ponder over the use of space, light and shadow, and you may find some triggering surprises or an entirely new approach to gardens.

Another and slightly less published area of garden design is rooftop gardens. In urban environment where buildings and roads have used up almost 90% of space that previously belonged to nature, we are taking first baby-steps back towards greener living and creating these little havens of plants and wilderness that so effectively calm us, revitalize. It is great to spot these little green patches from high above and to visualize what it is like to be within them. Seeing the empty, lonely roof turned into something wonderful, you will find immense satisfaction, will be excited about every plant that awakes in the spring and every single bloom that greets you. You will be quietly worried about some plants, some conditions that you did not expect, some little things gone wrong which is perfectly normal but overall, the feeling that you have created something that will be living its own life, evolving over time within restricted conditions, is truly exciting. I hope the section of roof gardens will inspire many to create more of these splendid spaces.

This book will encourage you to look at the world of gardens with a fresh perspective, igniting some great ideas, helps to turn some of your obstacles into stepping stones, and gives an opportunity to observe and understand, appreciating the ever-changing gardens of our time.

PREFACE 3

Since the days of pre-history, man has felt the need to manage the land around him, chiefly to provide food and shelter. Working on the land is now left to a minority. Nevertheless, it is still very much part of the human condition to need to connect with the natural world.

For many of us, this has led to an interest in the aesthetics of our surroundings and their ability to support our varied lifestyles. This is in the context of a time when the look of something is very important. As consumers, we are constantly bombarded by images of products we can buy for our homes, clothes we can wear, and cars we can drive.

In the past 50 years there has been a quiet revolution, especially for ordinary working people, in the feasibility of tailoring our homes to suit our requirements and personal taste. The employment of an interior designer or landscape architect used to be the preserve of only the wealthy, who had large houses and estates. Today, the means of transforming even the smallest garden of a modest suburban home is affordable and the numerous television programmes and magazine articles about home improvement encourage us to be adventurous and innovative. Thus the concept of garden design for all has been born.

Garden design is also strongly linked with our sense of identity. We want to put our stamp on our surroundings, firstly with the interior of our homes and now with our gardens. What does my garden say about me? How does it respond to my needs as an individual? Is it a fashion statement or a haven for wildlife? Can I grow herbs to compliment my cooking? Is it a safe playground for my children? Can I barbecue and entertain my friends? A garden is no longer just a place to hang out the washing or position the shed. It is an extension of the life we live inside our houses. It is a reflection of our personality, and our status, our interests.

Our towns and cities are getting ever more congested. We cry out for our piece of personal, private, green space. The garden can be our sanctuary; a place to re-energise; to re-connect with nature, in a frenetic, industrialised world. To be able to step outside our man-made structures into a living, breathing, constantly changing space, speaks to the soul.

What better job could a garden designer have than being part of this exhilarating movement?

Cherry Mills
Cherry Mills Garden Design

CONTENTS

The Best
Villa Garden Landscape

Garrell St

Landscape Architect
Steve Taylor

Firm
COS Design

Location
Caulfield, Australia

Photographer
Tim Turner

Briefly the owners wanted to create an outdoor space that was visually stunning from every internal vista that complimented the modern minimalistic architecture with a focus on entertaining.

The alfresco space offers wide open dining area with the warmth of the chunky yellow cedar timber arbor over and is direct extension of the home's interior living space. This space flows onto the court and offers access and space for the children to play. The angled cantilevered BBQ and built-in floating bench seating also adds functionality with the added element of architectural flair to the garden.

The front yard is framed by the unique architectural front fence and entry portico. Within the property the shadow lined anston blade wall doubles as a bench seat providing a quiet space for relaxation and overlooking the black mondo bands throughout the lush buffalo lawn. A matching aloe tree in a raised planter defines the entry and ties the front and rear zones together. The maple underplanted with gardenias breaks the mass of hard surfaces with cliveas being highlighted under the matching timber blades. The large format anston paving is broken up with thin strip bandings as a subtle highlight that is picked up by the black mondo strips.

In summary the pool and spa combo really stands out along the front fence/walls are the heroes of the space. In general, all elements work well together to create a warm overall grounded sophisticated end result.

OUTDOOR USE ONLY

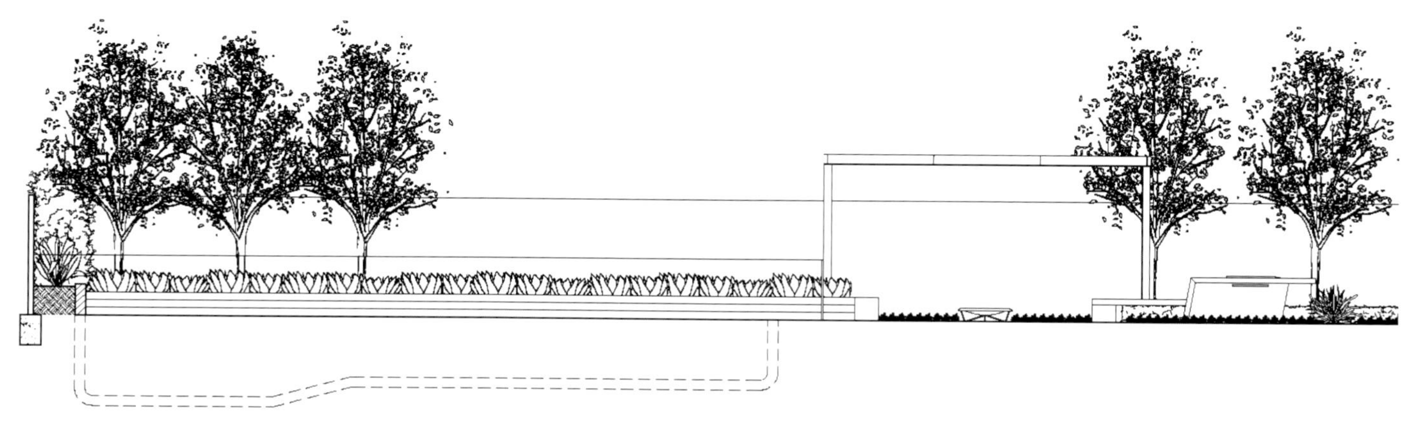

Poolside Features Elevation
SCALE 1: 100

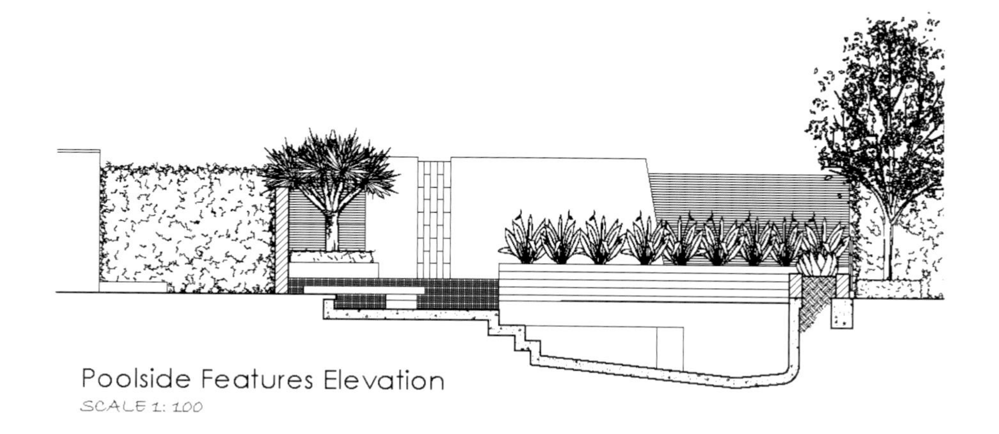

Poolside Features Elevation
SCALE 1: 100

Front Yard Elevation
SCALE 1: 100

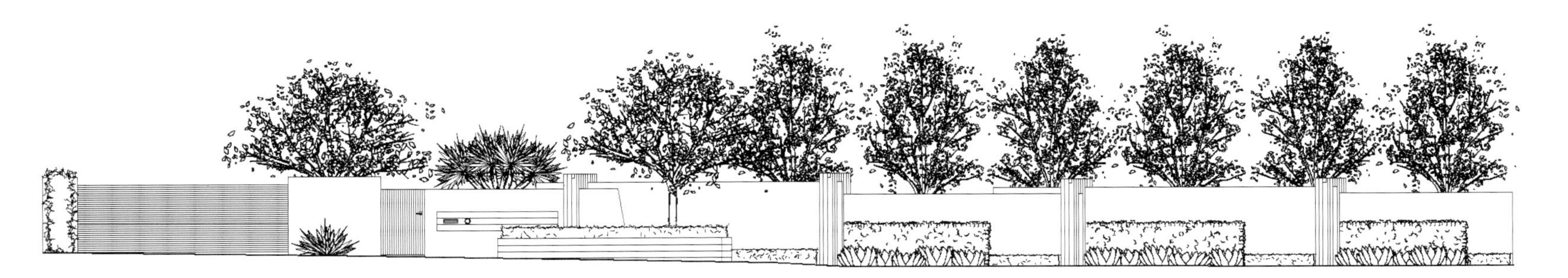

Western Streetscape & Pedestrian Entry Elevation
SCALE 1: 100

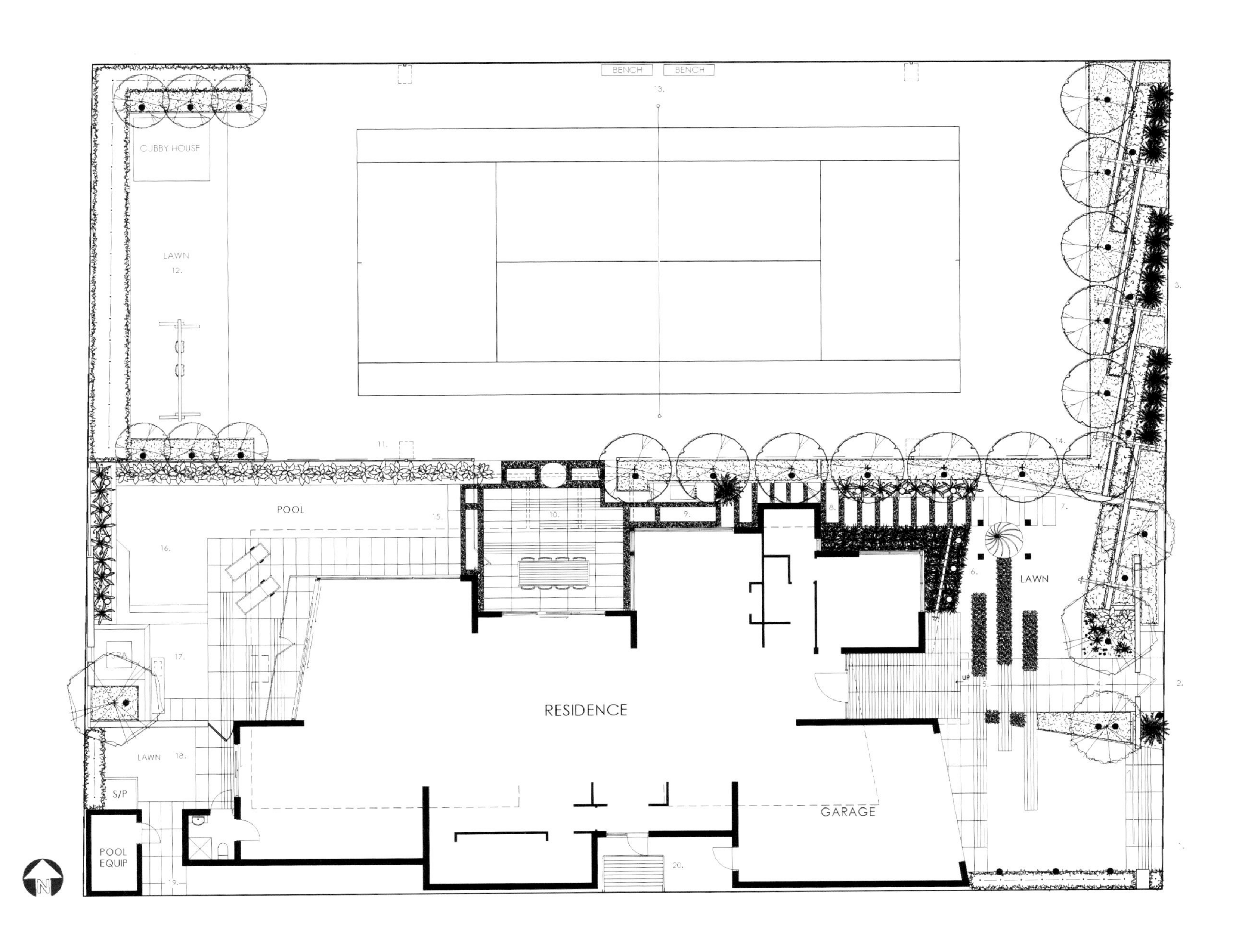

Proposed Lighting Specifications

●	31	MR16 35w / 38°	Directional Uplight. Black Powder-Coat Finish MEGABAY M2520
○	11	MR16 20w / 38°	Piccollo Uplight. Megachrome Finish MEGABAY M2600
▬	2	G4 20w BI-PIN	Nemo Step Light. Brushed Chrome Finish MEGABAY M2077
■	4	MR16 50w	Hunza Lawn Light Black LL BK

Parslow St

Landscape Architect
Steve Taylor

Firm
COS Design

Photographer
Tim Turner

The designers were approached by the clients for a new, more modern contemporary home which left very little space for garden. The approach and design principals needed to suit the new architecture and internal décor of the home. Although small, available space for design bears no relevance on the ease of the design process, to the contrary it complicates and makes the process more difficult as every square meter counts.

Faced with an imposing large garage backdrop in the rear courtyard meant a design challenge on the designers' hands. They needed to soften and minimise the focus of the garage structure within the small rear space. To achieve this lateral features were created to draw the focus of the eye away from the rear wall and to the sides of the courtyard. A large uncluttered open bluestone paved surface remained untouched and a custom three tiered water feature/reflection pond was created to the northern side of the courtyard. The water wall gives the water movement height and structure and creates a very subtle calming sound to the space while the random clad bluestone plinth creates a mirror like finish providing stunning reflective qualities of the Teddy Bear Magnolias (that line both of the courtyard boundaries), and the lower pond with its floating stepping stones cools the courtyard and gives the feature scale a presence within the space.

Projects of this quality and style don't just happen; they are only created through a highly creative professional, thorough design and construction processes. Steve and the team at COS Design offer these services to the highest level with their work being awarded multiple state and National awards.

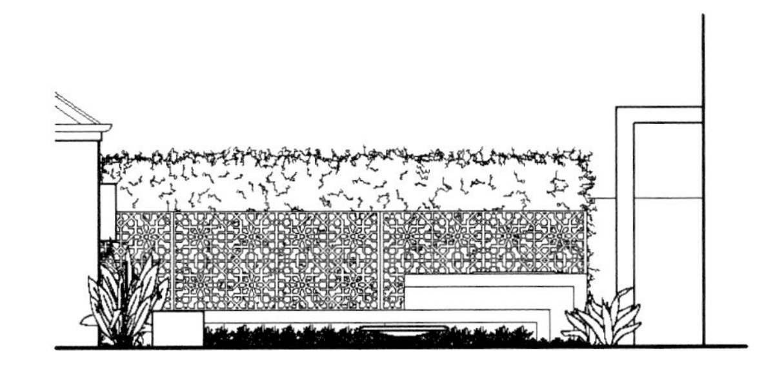

BBQ & Seating Elevation

SCALE 1: 100

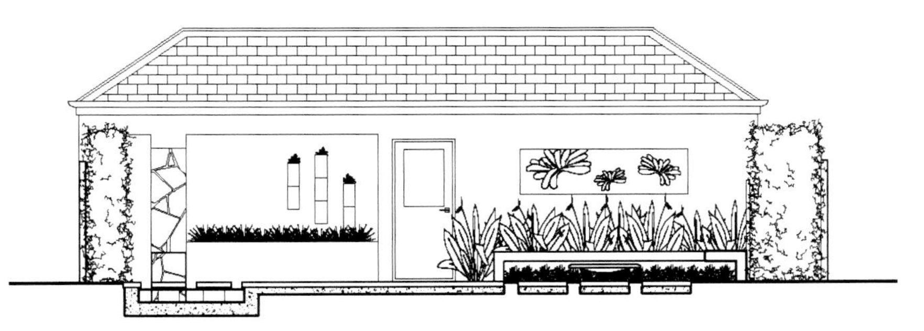

Garage & Water Feature Elevation

SCALE 1: 100

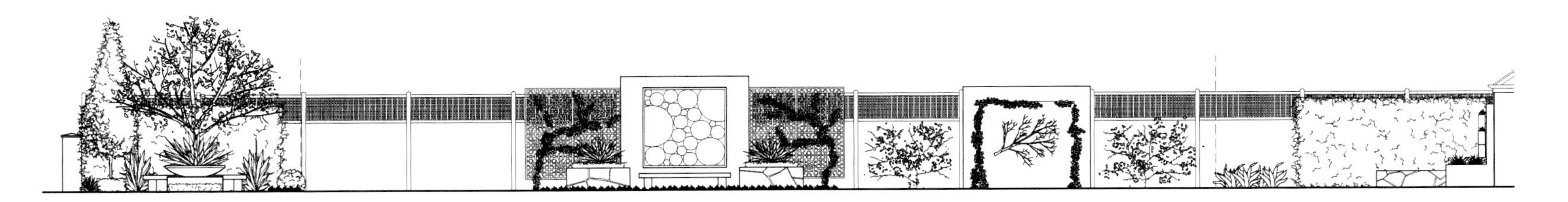

Northern Boundary Elevation with Visual Features

SCALE 1: 100

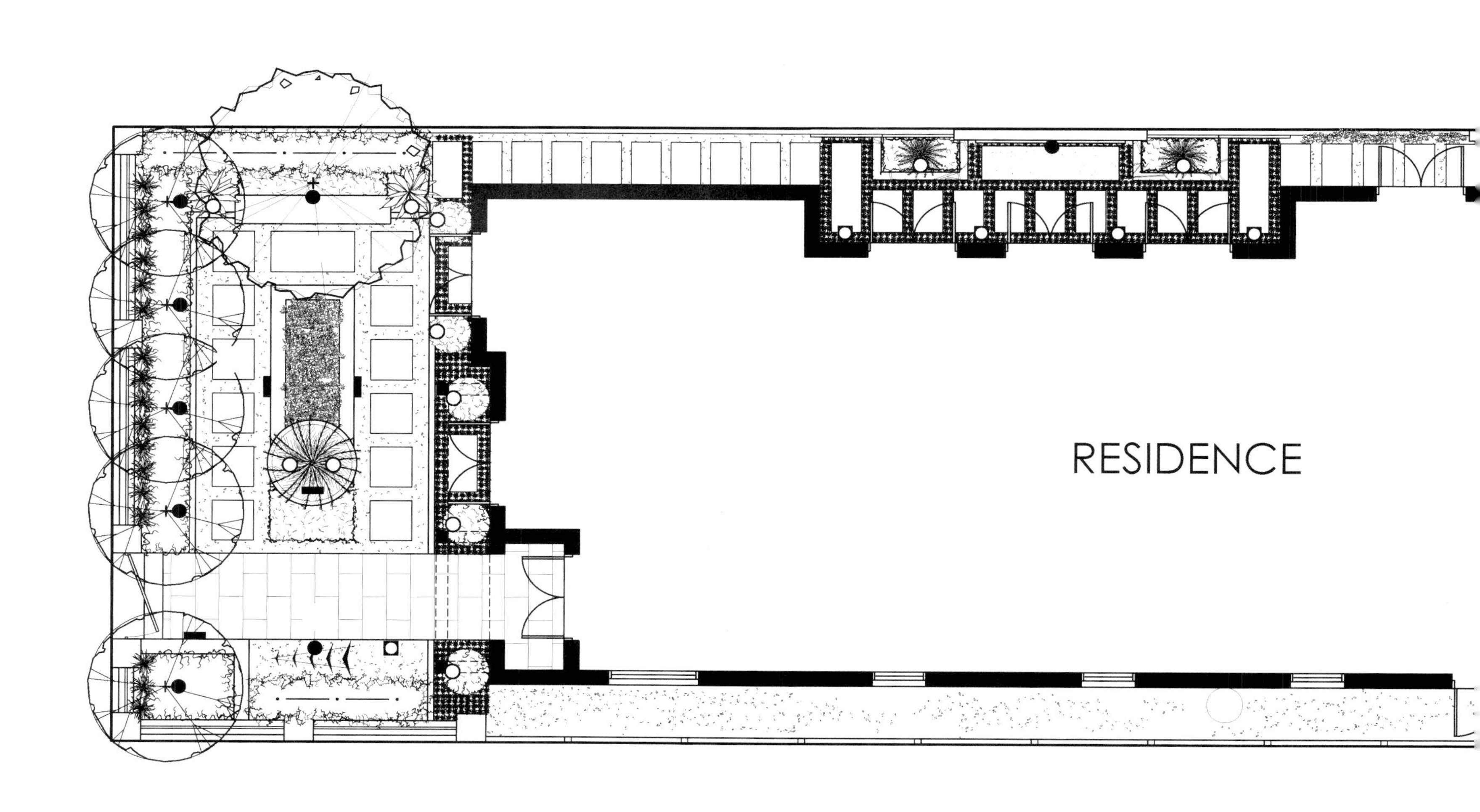
RESIDENCE

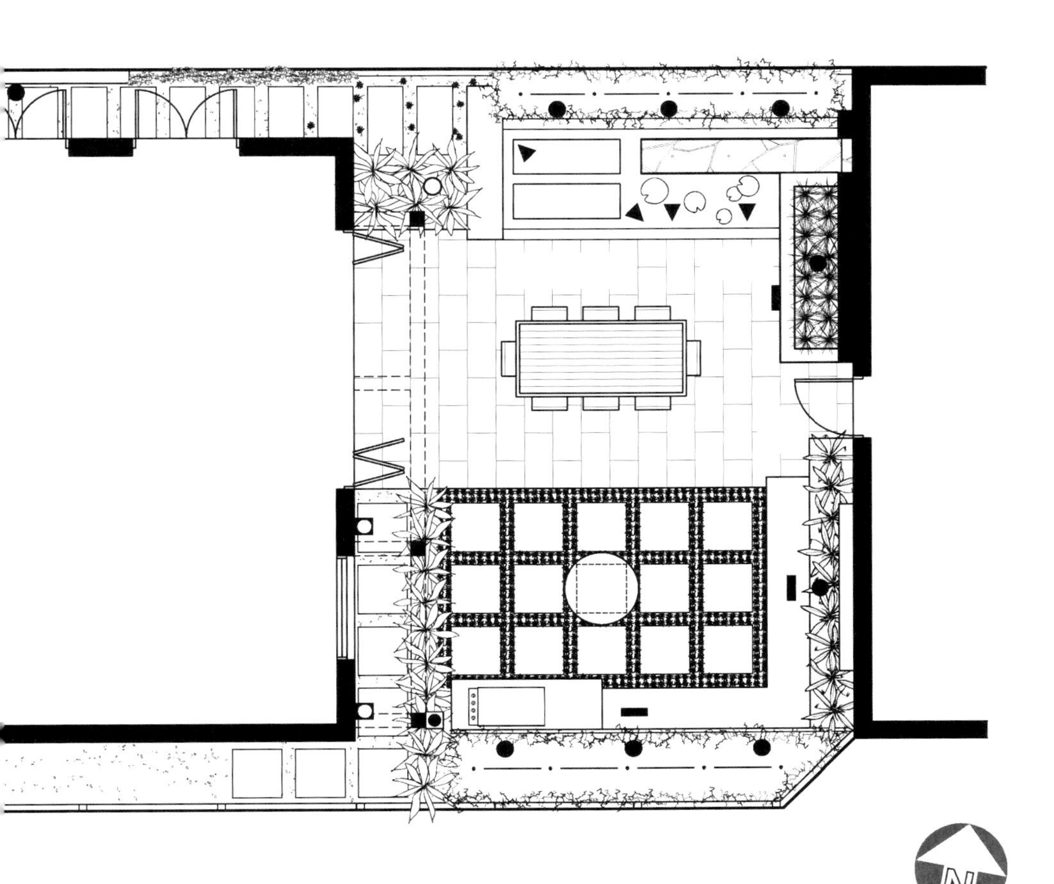
N

Radnor St

Landscape Architect
Steve Taylor

Firm
COS Design

Photographer
Tim Turner

Faced with the many challenges of a large imposing dwelling that dominated the streetscape and bare building site it inhabited, pre-set levels and infrastructure along with a slight conflict of house design to poolside pavilion design, the team set about to design a garden that would first and foremost soften the home from the streetscape, compliment the modern interior design and pool house architecture while still protecting the traditional facade of the home and create visually stunning, highly functional outdoor spaces.

The courtyards and water feature created exceed the client's expectations on every level. The water feature is the first to be created using LUMP's new 3D Sahara wall panelling and much research was done in relation to paint finishes and water proofing to protect the high density foam multi-panelled material. With LED strip lighting and up lighting on the water wall, dancing shadows fill the courtyard at night with shadows on the panel due to its 3D nature. The Himalayan weeping bamboo creates an architectural living sculpture in the space with the floating solid bluestone bench protruding from the spotted gum slatted wall. Floating steppers surrounded by black polished pebbles act as a dry extension of the child safe water wall.

Overall the garden exudes clever design principals, quality construction techniques and an overall sophisticated ambiance that the clients and the designer are proud of on every level.

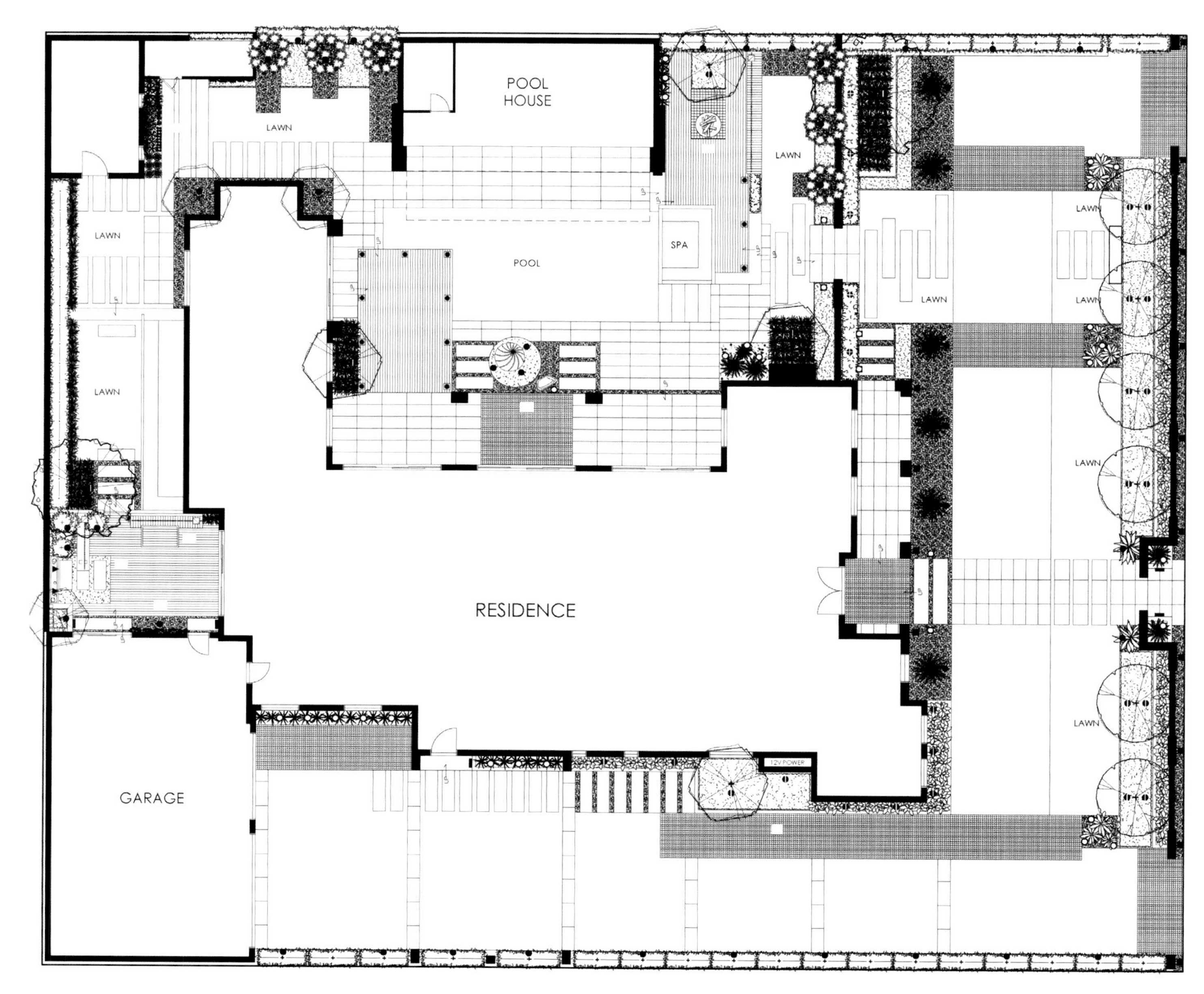

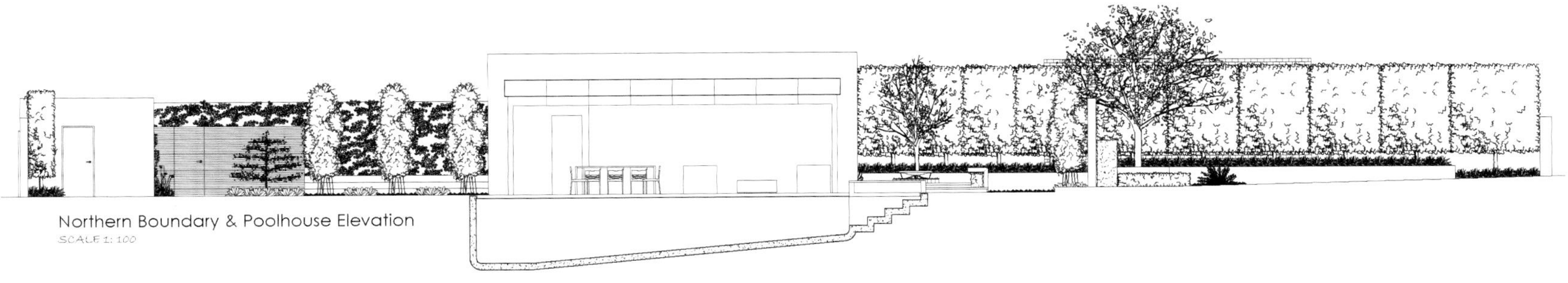

Northern Boundary & Poolhouse Elevation
SCALE 1: 100

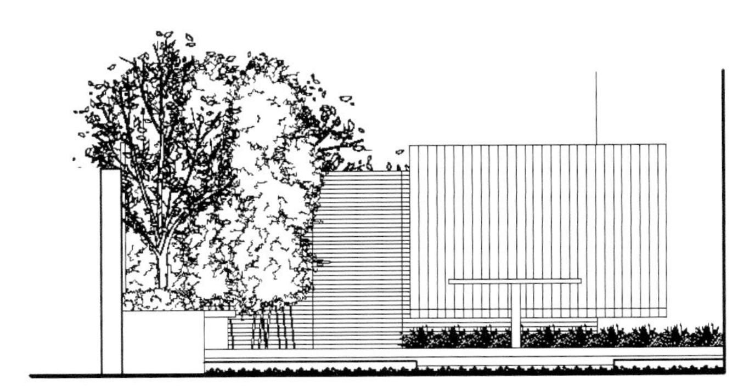

Northern Courtyard Elevation
SCALE 1: 100

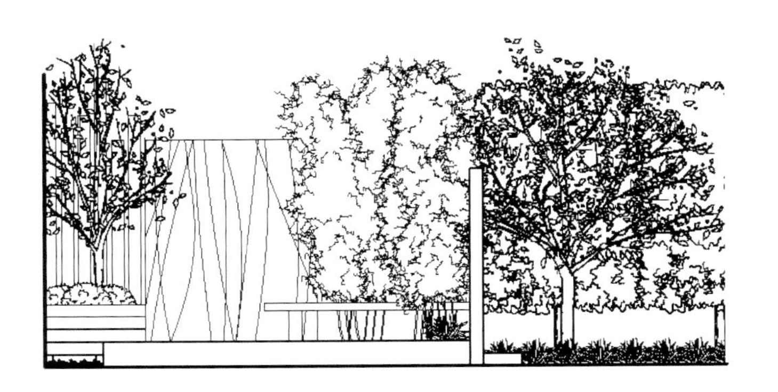

Western Courtyard Elevation
SCALE 1: 100

Water Garden Oasis

Landscape Architect
Lewis Aqüi Landscape + Architectural Design, LLC

Architect
Synergy Design Studio

Location
Miami Beach, Florida, USA

Photographer
Lewis Aqüi Photography

This residence was originally constructed decades ago on an Island just off Miami Beach, Florida. This extensive renovation included the creation of a contemporary, playful paradise for the client to use as a vacation home. The garden concept evolved to mimic the feeling of lounging in an upscale spa resort. The program included a motor court, an entry courtyard, a central water-garden 'oasis', a dramatically tiled swimming pool, stone terraces, a shady lounging pavilion, a therapeutic spa, an outdoor kitchen, and a gracious wood dock overlooking Biscayne Bay.

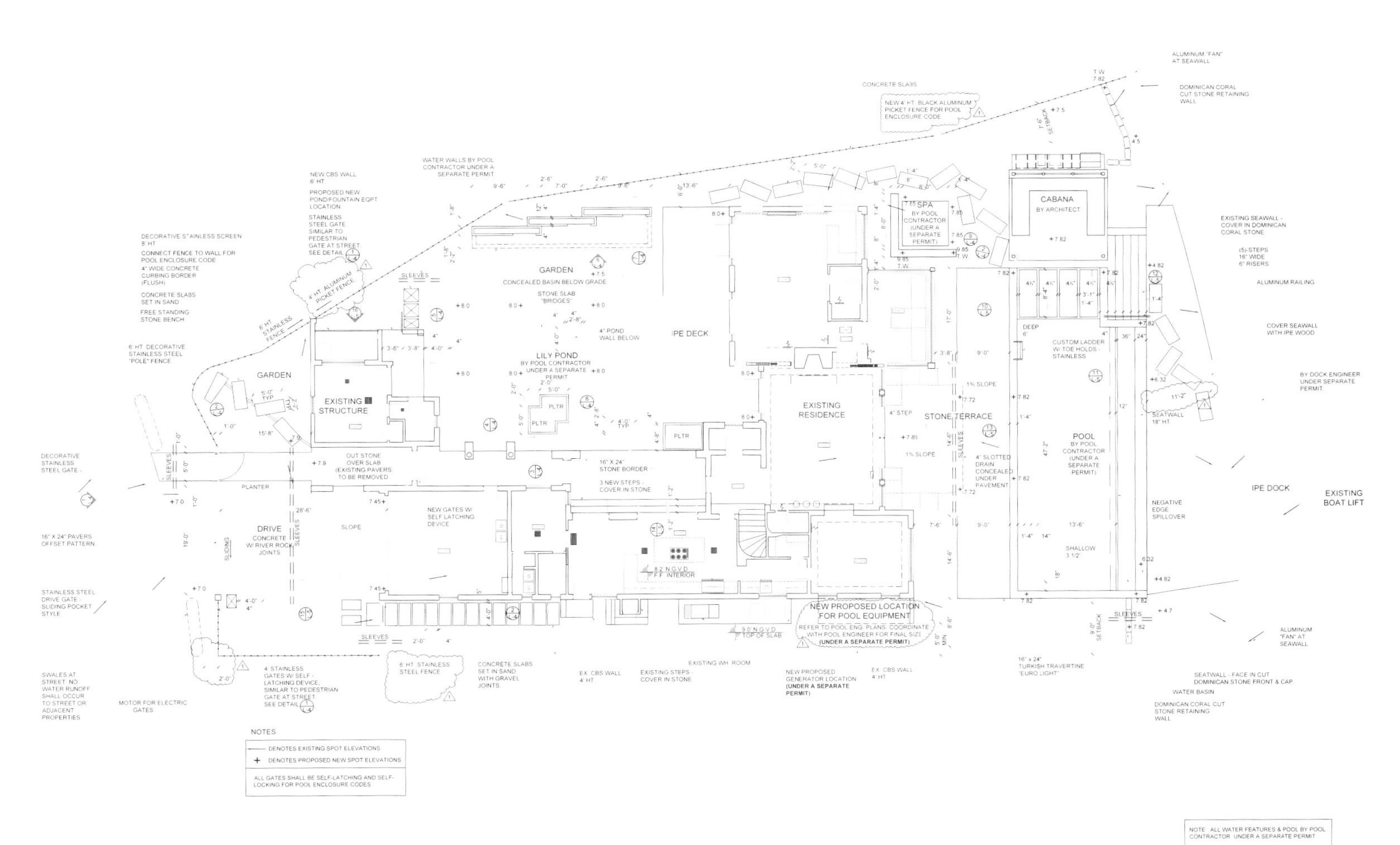
CABANA
BY ARCHITECT
SPA
GARDEN
CONCEALED BASIN BELOW GRADE
LILY POND
IPE DECK
EXISTING STRUCTURE
GARDEN
EXISTING RESIDENCE
STONE TERRACE
POOL
IPE DOCK
EXISTING BOAT LIFT
DRIVE
NEW PROPOSED LOCATION FOR POOL EQUIPMENT
(UNDER A SEPARATE PERMIT)
NOTES
DENOTES EXISTING SPOT ELEVATIONS
DENOTES PROPOSED NEW SPOT ELEVATIONS
ALL GATES SHALL BE SELF-LATCHING AND SELF-LOCKING FOR POOL ENCLOSURE CODES
NOTE: ALL WATER FEATURES & POOL BY POOL CONTRACTOR UNDER A SEPARATE PERMIT
SITE PLAN

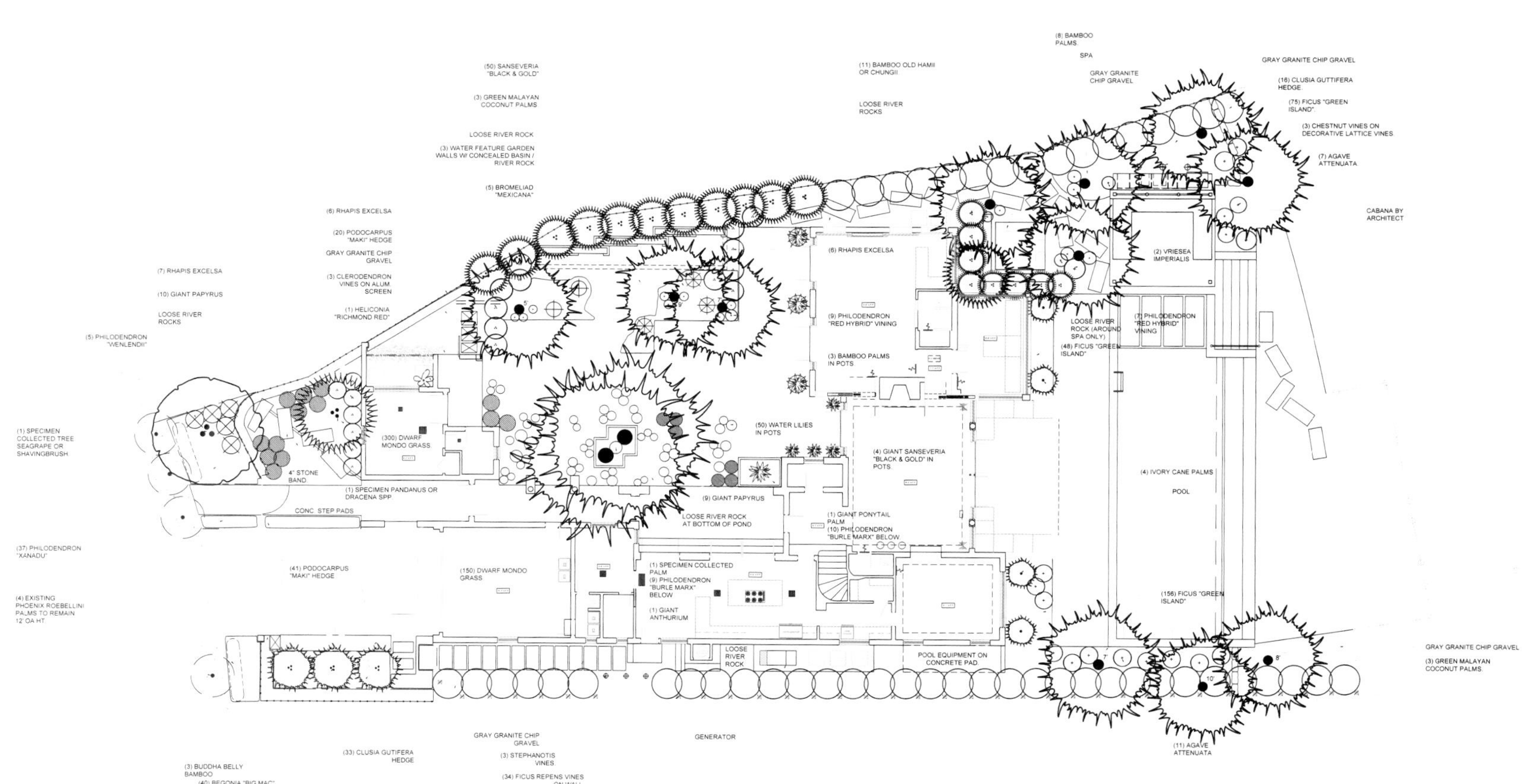
(50) SANSEVERIA "BLACK & GOLD"
(3) GREEN MALAYAN COCONUT PALMS
LOOSE RIVER ROCK
(3) WATER FEATURE GARDEN WALLS W/ CONCEALED BASIN / RIVER ROCK
(5) BROMELIAD "MEXICANA"
(6) RHAPIS EXCELSA
(20) PODOCARPUS "MAKI" HEDGE
GRAY GRANITE CHIP GRAVEL
(3) CLERODENDRON VINES ON ALUM. SCREEN
(1) HELICONIA "RICHMOND RED"
(7) RHAPIS EXCELSA
(10) GIANT PAPYRUS
LOOSE RIVER ROCKS
(5) PHILODENDRON "WENLENDII"
(1) SPECIMEN COLLECTED TREE SEAGRAPE OR SHAVINGBRUSH
4" STONE BAND
(300) DWARF MONDO GRASS
(1) SPECIMEN PANDANUS OR DRACENA SPP.
CONC. STEP PADS
(37) PHILODENDRON "XANADU"
(41) PODOCARPUS "MAKI" HEDGE
(150) DWARF MONDO GRASS
(4) EXISTING PHOENIX ROEBELLINI PALMS TO REMAIN 12' OA HT.
(3) BUDDHA BELLY BAMBOO
(40) BEGONIA "BIG MAC"
(33) CLUSIA GUTIFERA HEDGE
GRAY GRANITE CHIP GRAVEL
(3) STEPHANOTIS VINES
(34) FICUS REPENS VINES ON WALL
GENERATOR
(11) BAMBOO OLD HAMII OR CHUNGII
LOOSE RIVER ROCKS
(8) BAMBOO PALMS
SPA
GRAY GRANITE CHIP GRAVEL
(6) RHAPIS EXCELSA
(9) PHILODENDRON "RED HYBRID" VINING
(3) BAMBOO PALMS IN POTS
(50) WATER LILIES IN POTS
(4) GIANT SANSEVERIA "BLACK & GOLD" IN POTS
(9) GIANT PAPYRUS
LOOSE RIVER ROCK AT BOTTOM OF POND
(1) GIANT PONYTAIL PALM
(10) PHILODENDRON "BURLE MARX" BELOW
(1) SPECIMEN COLLECTED PALM
(9) PHILODENDRON "BURLE MARX" BELOW
(1) GIANT ANTHURIUM
LOOSE RIVER ROCK
POOL EQUIPMENT ON CONCRETE PAD
GRAY GRANITE CHIP GRAVEL
(16) CLUSIA GUTTIFERA HEDGE
(75) FICUS "GREEN ISLAND"
(3) CHESTNUT VINES ON DECORATIVE LATTICE VINES
(7) AGAVE ATTENUATA
CABANA BY ARCHITECT
(2) VRIESEA IMPERIALIS
LOOSE RIVER ROCK (AROUND SPA ONLY)
(48) FICUS "GREEN ISLAND"
(7) PHILODENDRON "RED HYBRID" VINING
(4) IVORY CANE PALMS
POOL
(156) FICUS "GREEN ISLAND"
9'
7'
5'
8'
10'
(11) AGAVE ATTENUATA
GRAY GRANITE CHIP GRAVEL
(3) GREEN MALAYAN COCONUT PALMS
SITE PLAN
SCALE: 1/8" = 1'-0"

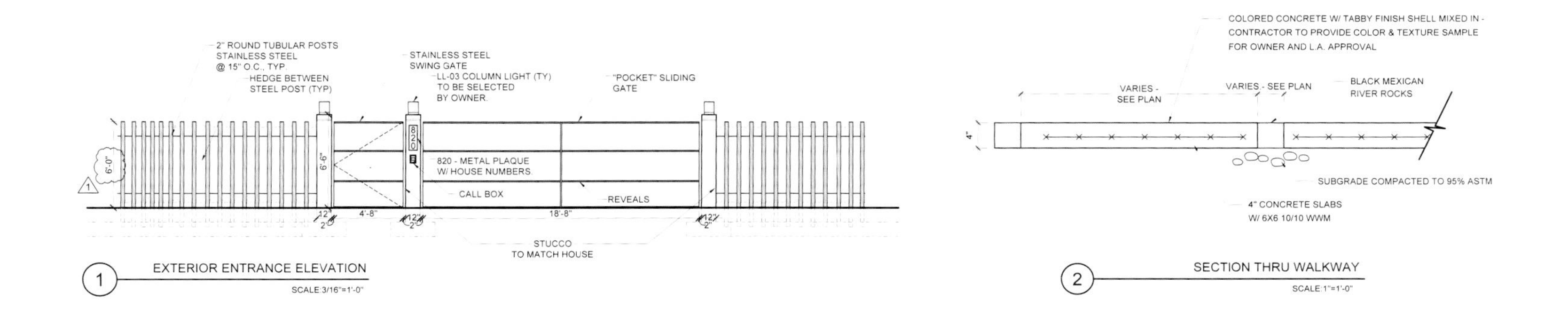

1 EXTERIOR ENTRANCE ELEVATION

SCALE 3/16"=1'-0"

2 SECTION THRU WALKWAY

SCALE 1"=1'-0"

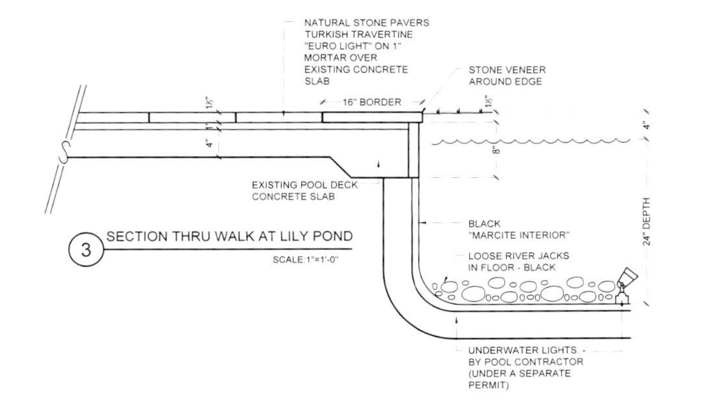

3 SECTION THRU WALK AT LILY POND

SCALE 1"=1'-0"

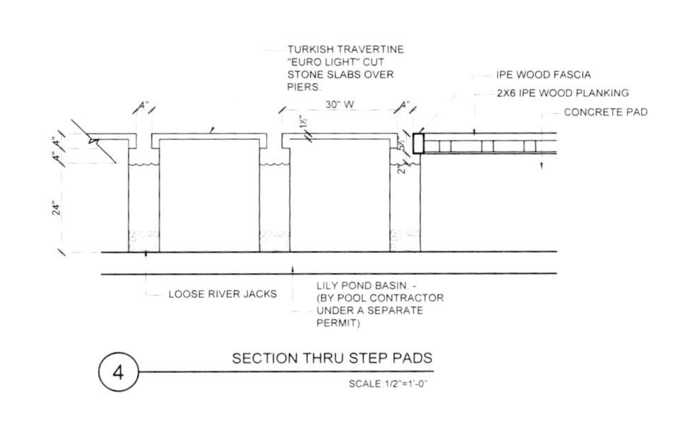

4 SECTION THRU STEP PADS

SCALE 1/2"=1'-0"

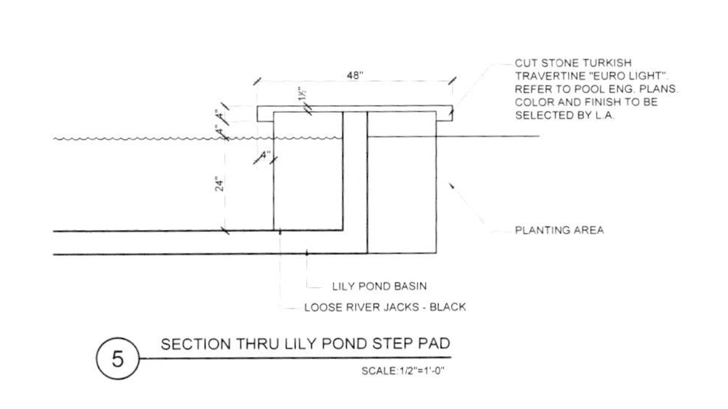

5 SECTION THRU LILY POND STEP PAD

SCALE 1/2"=1'-0"

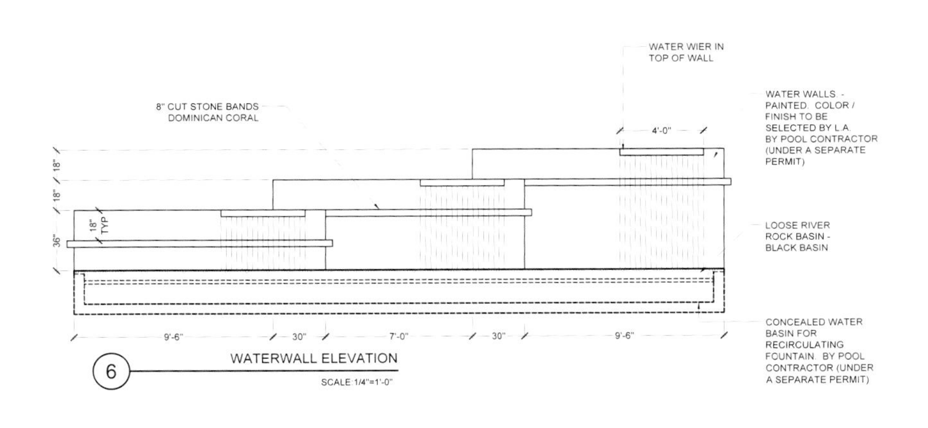

6 WATERWALL ELEVATION

SCALE 1/4"=1'-0"

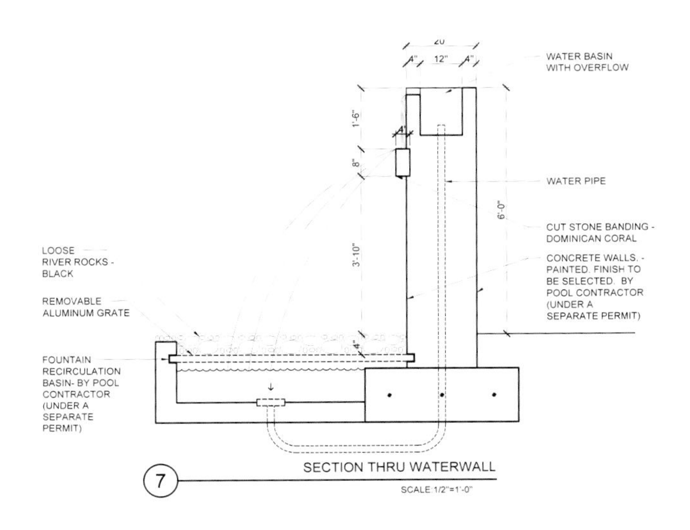

7 SECTION THRU WATERWALL

SCALE 1/2"=1'-0"

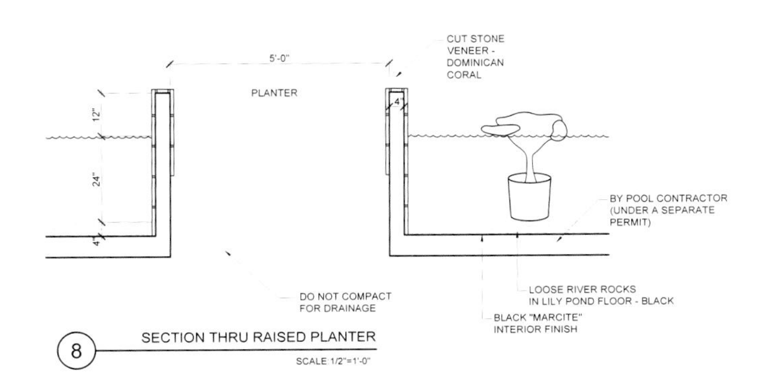

8 SECTION THRU RAISED PLANTER

SCALE 1/2"=1'-0"

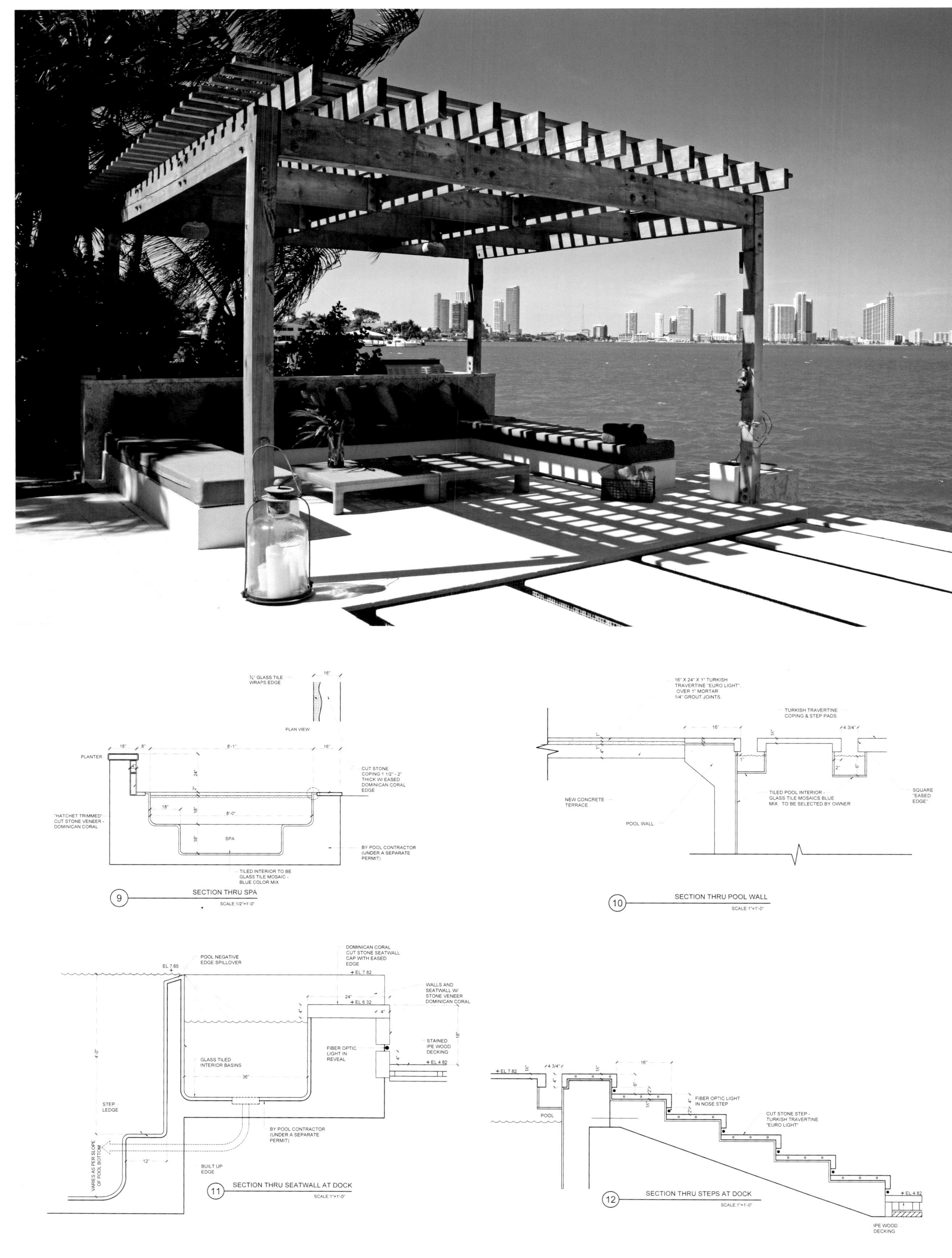

¾" GLASS TILE WRAPS EDGE
PLAN VIEW
PLANTER
CUT STONE COPING 1 1/2" - 2" THICK W/ EASED DOMINICAN CORAL EDGE
"HATCHET TRIMMED" CUT STONE VENEER - DOMINICAN CORAL
SPA
BY POOL CONTRACTOR (UNDER A SEPARATE PERMIT)
TILED INTERIOR TO BE GLASS TILE MOSAIC - BLUE COLOR MIX
9
SECTION THRU SPA
SCALE:1/2"=1'-0"
16" X 24" X 1" TURKISH TRAVERTINE "EURO LIGHT", OVER 1" MORTAR. 1/4" GROUT JOINTS.
TURKISH TRAVERTINE COPING & STEP PADS.
NEW CONCRETE TERRACE.
POOL WALL
TILED POOL INTERIOR - GLASS TILE MOSAICS BLUE MIX. TO BE SELECTED BY OWNER
SQUARE "EASED EDGE"
10
SECTION THRU POOL WALL
SCALE:1"=1'-0"
POOL NEGATIVE EDGE SPILLOVER
EL 7.65
DOMINICAN CORAL CUT STONE SEATWALL CAP WITH EASED EDGE
+ EL 7.82
WALLS AND SEATWALL W/ STONE VENEER DOMINICAN CORAL
+ EL 6.32
FIBER OPTIC LIGHT IN REVEAL
STAINED IPE WOOD DECKING
+ EL 4.82
GLASS TILED INTERIOR BASINS
STEP LEDGE
BY POOL CONTRACTOR (UNDER A SEPARATE PERMIT)
VARIES AS PER SLOPE OF POOL BOTTOM
BUILT UP EDGE
11
SECTION THRU SEATWALL AT DOCK
SCALE:1"=1'-0"
+ EL 7.82
POOL
FIBER OPTIC LIGHT IN NOSE STEP
CUT STONE STEP - TURKISH TRAVERTINE "EURO LIGHT"
+ EL 4.82
IPE WOOD DECKING
12
SECTION THRU STEPS AT DOCK
SCALE:1"=1'-0"

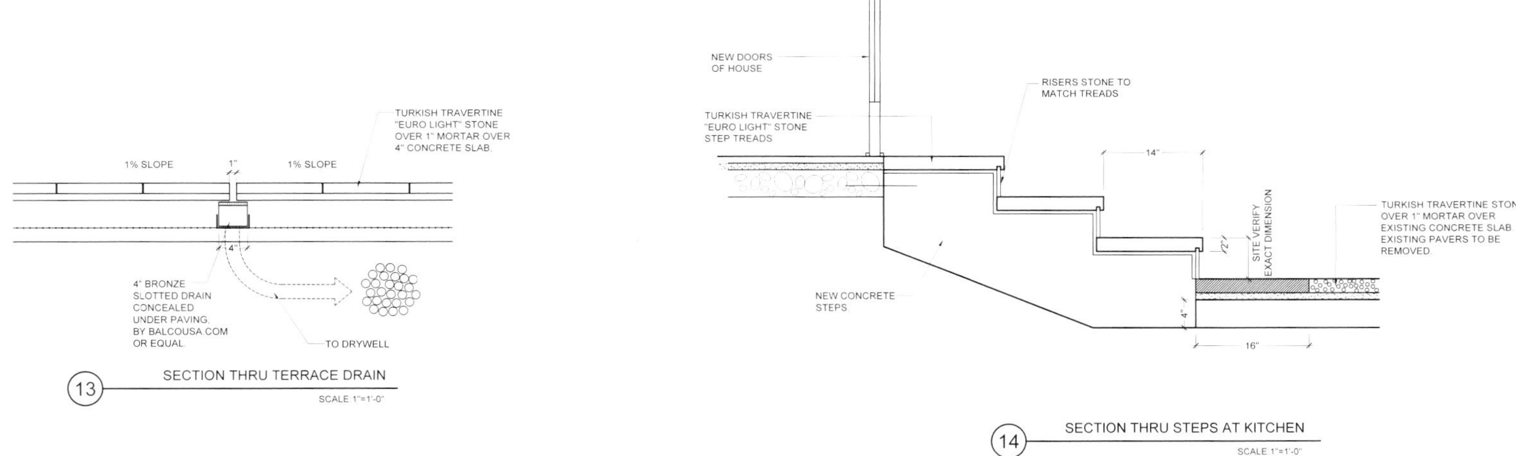

13 SECTION THRU TERRACE DRAIN
SCALE 1"=1'-0"

14 SECTION THRU STEPS AT KITCHEN
SCALE 1"=1'-0"

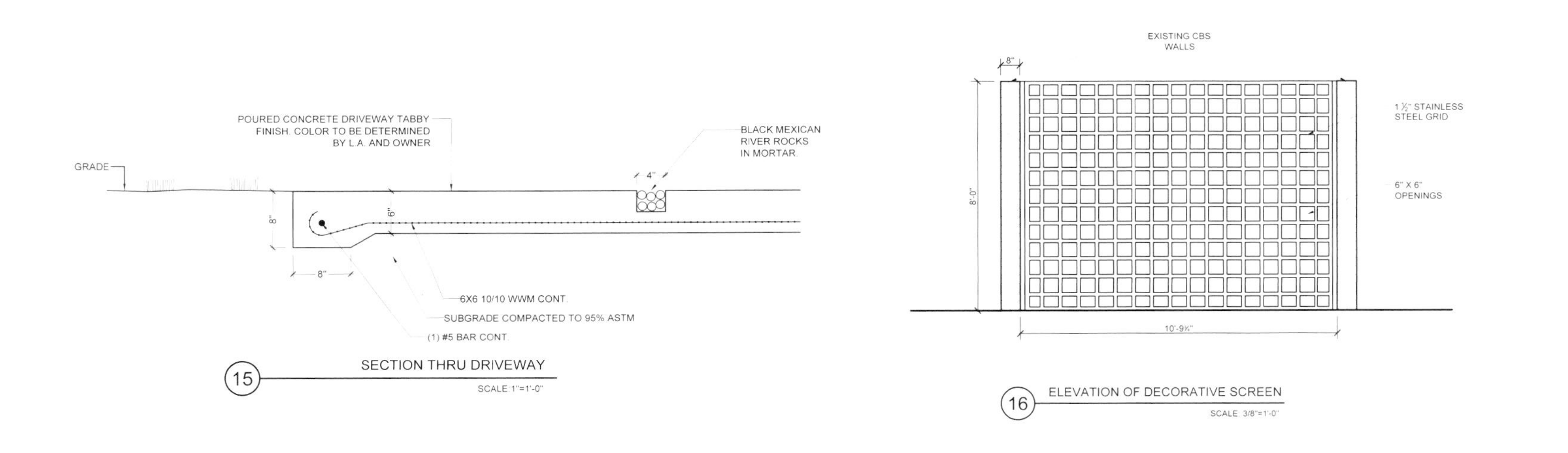

15 SECTION THRU DRIVEWAY
SCALE 1"=1'-0"

16 ELEVATION OF DECORATIVE SCREEN
SCALE 3/8"=1'-0"

Melbourne International Flower & Garden Show 2007

Landscape Architect
Dean Herald

Location
Melbourne, Victoria, Australia

Landscape Construction
Rolling Stone Landscapes

Photographer
Danny Kildare

The Melbourne International Flower & Garden show is not only Australia's largest garden show, it is a world renowned event in the landscape and gardening industry showcasing designer gardens with the latest in outdoor design trends. The brief the Dean set was the concept of 'Resort Style Living' to express the level of function, fun and relaxation that can be had in a modern-day garden.

A fully tiled pool acts as the centre piece of the garden with which all other spaces interact. The pool features a swim-up wet bar that includes a fridge, sink & storage.

Adjacent to the pool is a cooking and dining pavilion which features a multipurpose handcrafted floating kitchen. A five metre long Corian counter is suspended from the ceiling and doubles as a dining table one end, cooking station the other.

On the lower level next to the pavilion, a timber deck forms the lounge area which has been positioned to capture a view of the pool including the cascading water wet edge.

The materials used throughout the design are natural stones & timbers which bring warmth and texture to the space. A stainless steel sphere sculpture is positioned in the higher pool pond as a focal point linking the other stainless steel items installed within the garden.

The planting used throughout the garden demonstrates contrasting texture, colour and movement. All provide interest and work well with the theme and colouring of the garden, enhancing the resort style feel.

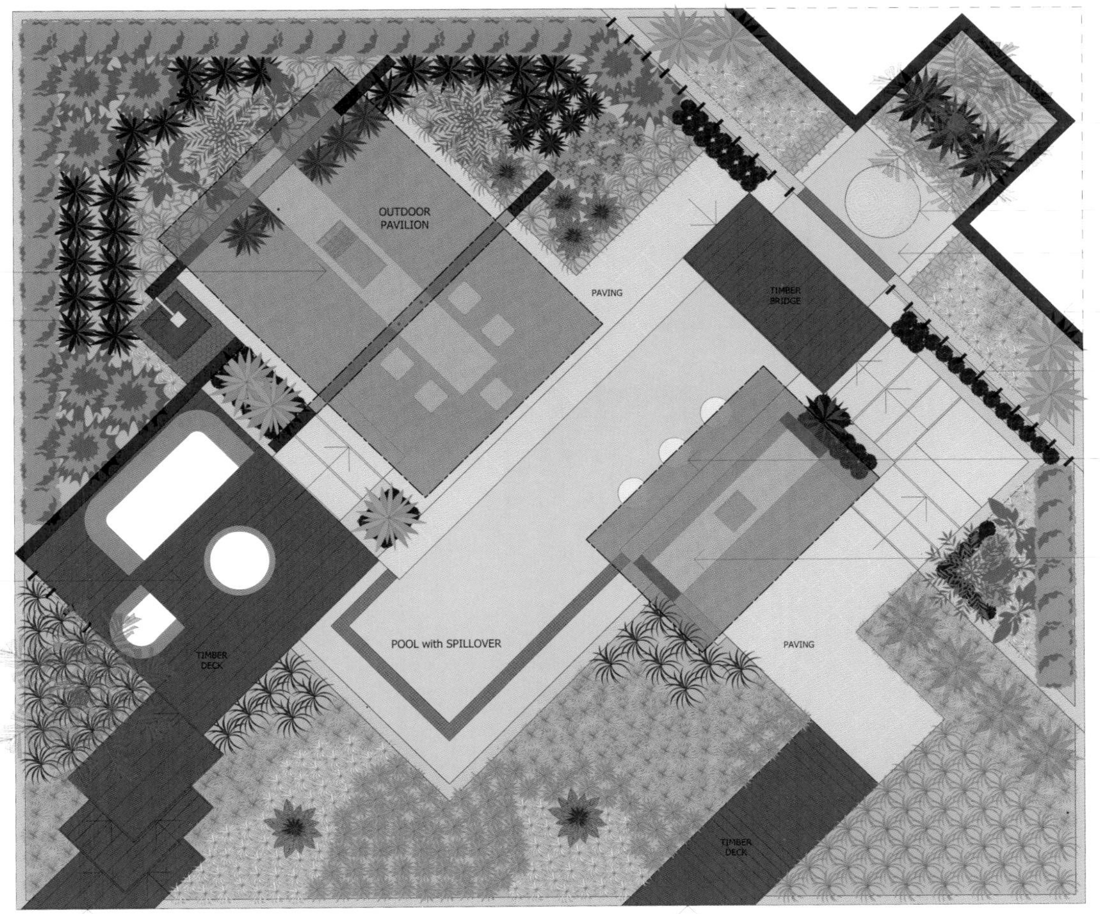
CANTILEVERED BENCH with BBQ & DINING AREA
OUTDOOR SHOWER
SUNKEN LOUNGE AREA
OUTDOOR PAVILION
PAVING
TIMBER BRIDGE
POOL with SPILLOVER
TIMBER DECK
PAVING
TIMBER DECK
DRAGONS BLOOD FEATURE TREE
STAINLESS STEEL SCULPTURE
UPPER POND with SPILLOVER
STONE CLAD WALL
TIMBER FINNS
SWIM UP BAR with STAINLESS STEEL STOOLS
SUNKEN BAR with SINK & FRIDGE

Crown

Melbourne International Flower & Garden Show 2011

Landscape Architect
Dean Herald

Location
Melbourne, Victoria, Australia

Landscape Construction
Rolling Stone Landscapes

Photographer
Danny Kildare

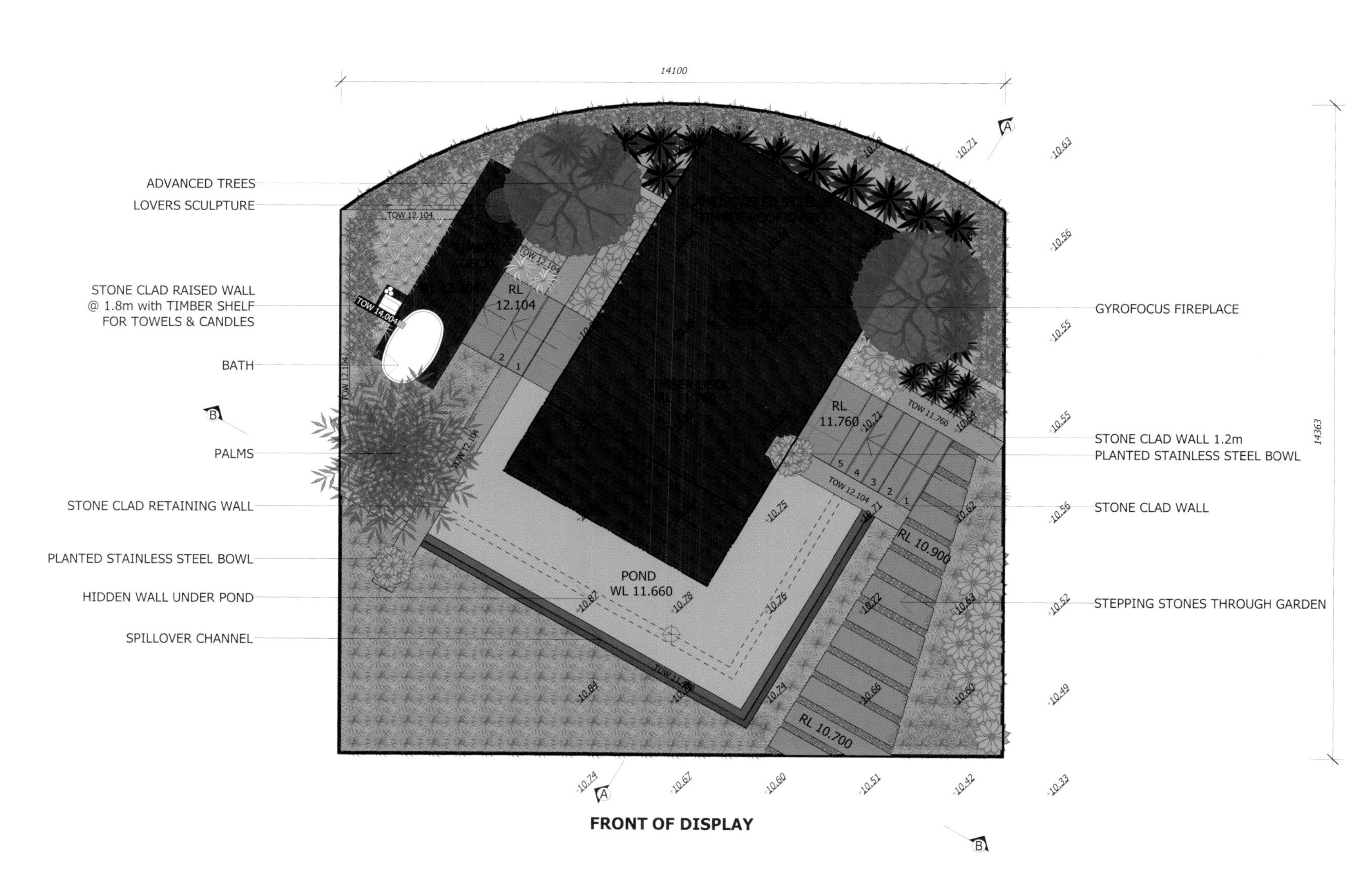

Returning to the Melbourne International Flower and Garden Show for the third time, Dean saw this as an important opportunity to send a message through a designed space that had real meaning and could assist in the greater awareness of a cause. Drawn to the effects that the different forms of mental illness were having on the community, Dean used this as an inspiration to form the design.

The look and feel of the garden has been designed to be one of a relaxing atmosphere whilst still having a structure of strength as the centre piece. The colours, textures and plant palette have all been selected to play a role in this along with the scale of individual items.

The stepping-stone path at the entry of the garden represents the beginning of the journey. At the end of the path, a wall encourages one to turn and take the stairs up to a new space. This represents a poignant moment in the healing process – the 'turning point' for those who suffer – and celebrates the moment they make the decision to take action towards a new direction in their lives.

At the top of the stairs is a space where key elements reside and contribute to the message of the garden. Fire and water, two essential elements for survival, are included whilst a roof structure with a 180-degree curve and large scale represents the milestone of 'turning your life around'.

A timber deck with a simple setting of two chairs provides a place of reflection & connection. Surrounding this space is a pond, along with a lush planting palette to soften the structures, as if softening the harshness of the illness or providing a sanctuary from it.

Above this area is the outdoor bathroom, another space of reflection and healing, providing a unique experience in the garden.

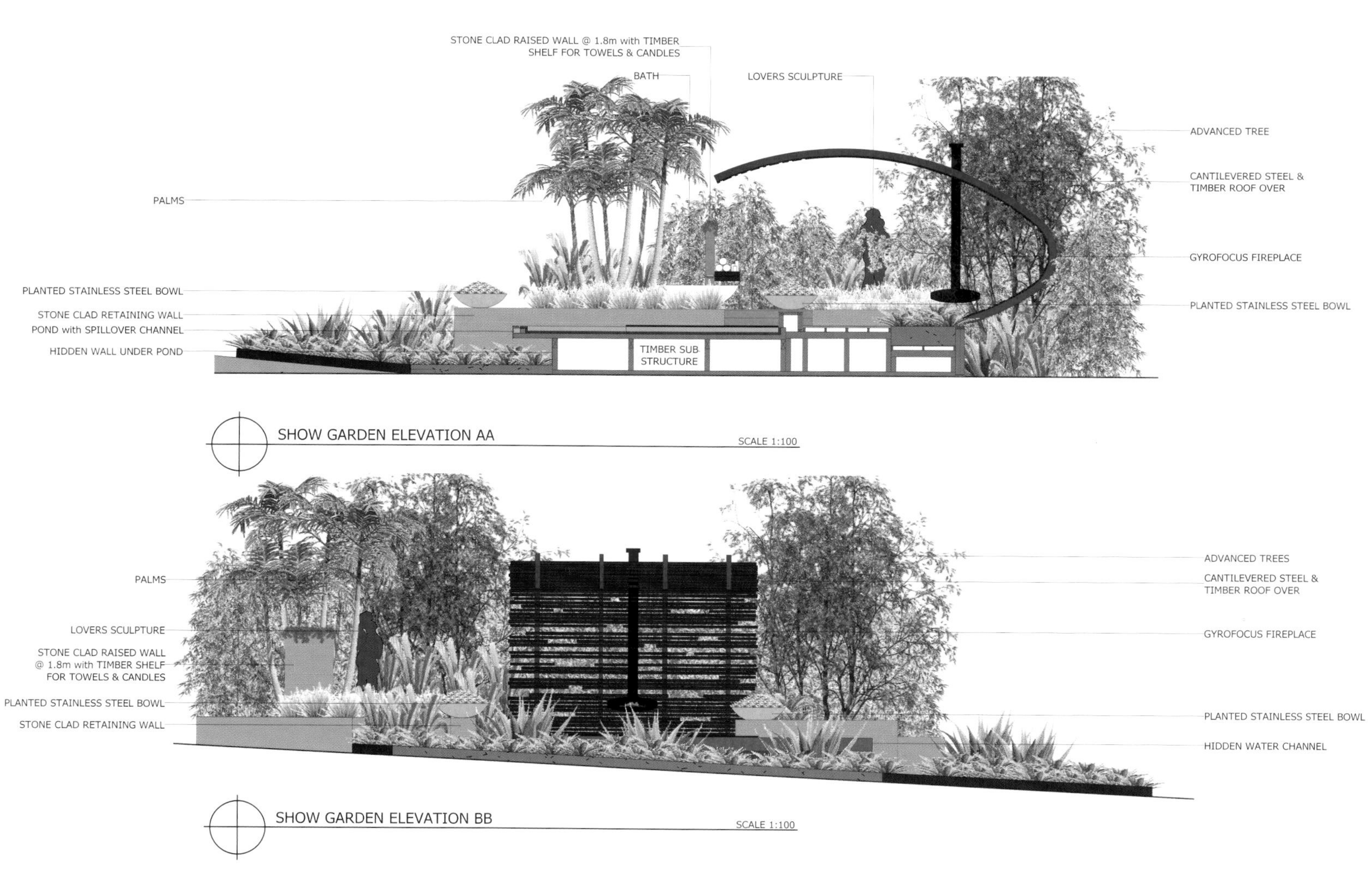

STONE CLAD RAISED WALL @ 1.8m with TIMBER SHELF FOR TOWELS & CANDLES
BATH
LOVERS SCULPTURE
ADVANCED TREE
CANTILEVERED STEEL & TIMBER ROOF OVER
PALMS
GYROFOCUS FIREPLACE
PLANTED STAINLESS STEEL BOWL
STONE CLAD RETAINING WALL
POND with SPILLOVER CHANNEL
HIDDEN WALL UNDER POND
PLANTED STAINLESS STEEL BOWL
TIMBER SUB STRUCTURE
SHOW GARDEN ELEVATION AA
SCALE 1:100
ADVANCED TREES
PALMS
CANTILEVERED STEEL & TIMBER ROOF OVER
LOVERS SCULPTURE
GYROFOCUS FIREPLACE
STONE CLAD RAISED WALL @ 1.8m with TIMBER SHELF FOR TOWELS & CANDLES
PLANTED STAINLESS STEEL BOWL
PLANTED STAINLESS STEEL BOWL
STONE CLAD RETAINING WALL
HIDDEN WATER CHANNEL
SHOW GARDEN ELEVATION BB
SCALE 1:100

Point Piper 'Sauvage'

Landscape Architect
ASPECT Studios

Architect
Tzannes Associates

Other Consultants
N Scape Landscape Professionals

Main Contractor
Earthtone Landscape Construction

Location
Sydney, New South Wales, Australia

Area
368 m²

Photographer
Simon Wood

The unique brief was that the spectacular garden overlooking Sydney Harbour should embody the notion of the 'Sauvage' (which is French for savage or wild).

The landscape needed to be a range of things including informal, relaxed, bold, eclectic and expressive of its genius loci.

To achieve the notion of the 'Sauvage', the designers installed planting at densities which result in vigorous and lush coverage, particularly in the sheltered areas. The design creates a sense of connected garden rooms on this multilevel site.

A stacked blue-grey stone wall defines the eastern edge of the driveway and wraps around to the front of the property. Thick slabs of basalt form the main staircase from the driveway up to the entry terrace. An in-ground timber sleeper deck in front of the rear entry door of the house provides an outdoor dining area amongst the greenery.

The gardens work perfectly in sympathy with the architecture enhancing the distinctive curve of the house and the large terrace overlooking Sydney Harbour.

The careful plant selection suits site conditions and the detailing resolve level changes, creating consistency and flow around the house.

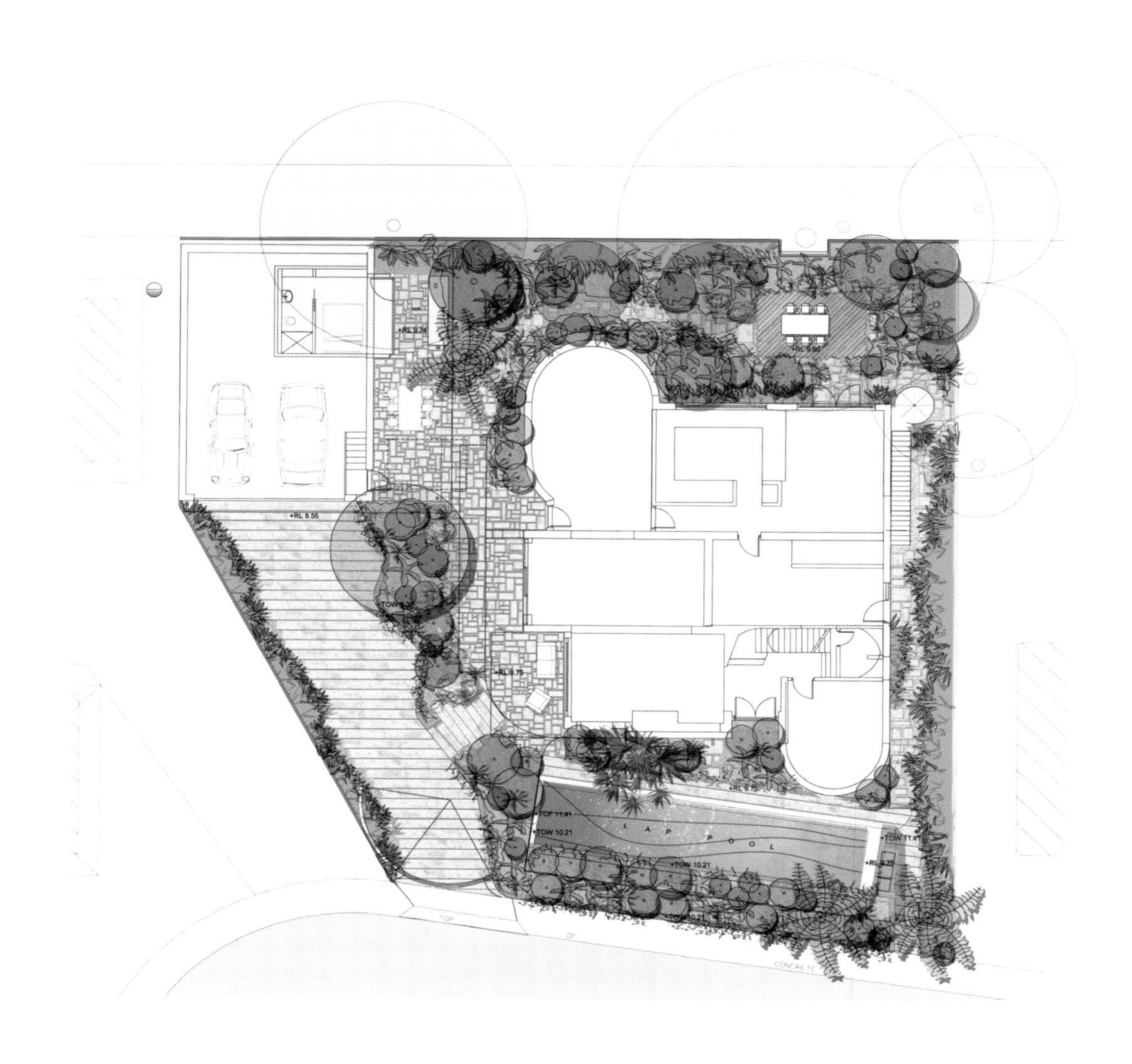

S. K. House Çubuklu Valley

Landscape Architect
Deniz Aslan, Selin Etkinöz Tunçer

Firm
DS Architecture – Landscape

Location
Beykoz, Istanbul, Turkey

Area
470 m²

Photographer
Gürkan Akay

S. K. House, located in Çubuklu Valley, presented with the opportunity to work on the landscape design of a mid-size house. The house was situated on a very steep site with different levels.

The project was divided into parts of an entrance garden, a slope garden and a lower garden. For the entrance, a peaceful, introverted, playful garden was designed. A reflection pool leads the way to this courtyard garden. The stones on the floor turn into seats and tables at some spots or are cleared out to provide space for a tree. This courtyard is an indoor garden that works with the elevation differences. Its background is a gramineae and bamboo garden ending with Cor-ten steel platforms. Paying attention not to harm the façade, the garden ends with simple details and a swing facing the slope.

The slope garden represents the existing rocky ground and its natural flora of pseudo-maquis. It follows the façade to the lower garden.

The lower garden unites the wooden decks going along the façade with a water surface. This is a water garden. The water element was approached as a swimming pool and a vegetational pool. It is designed as an infinity pool extending towards the Bosphorus. A gazebo placed on the platform separating the swimming and vegetational areas offers a view of Istanbul in a micro-climatic environment.

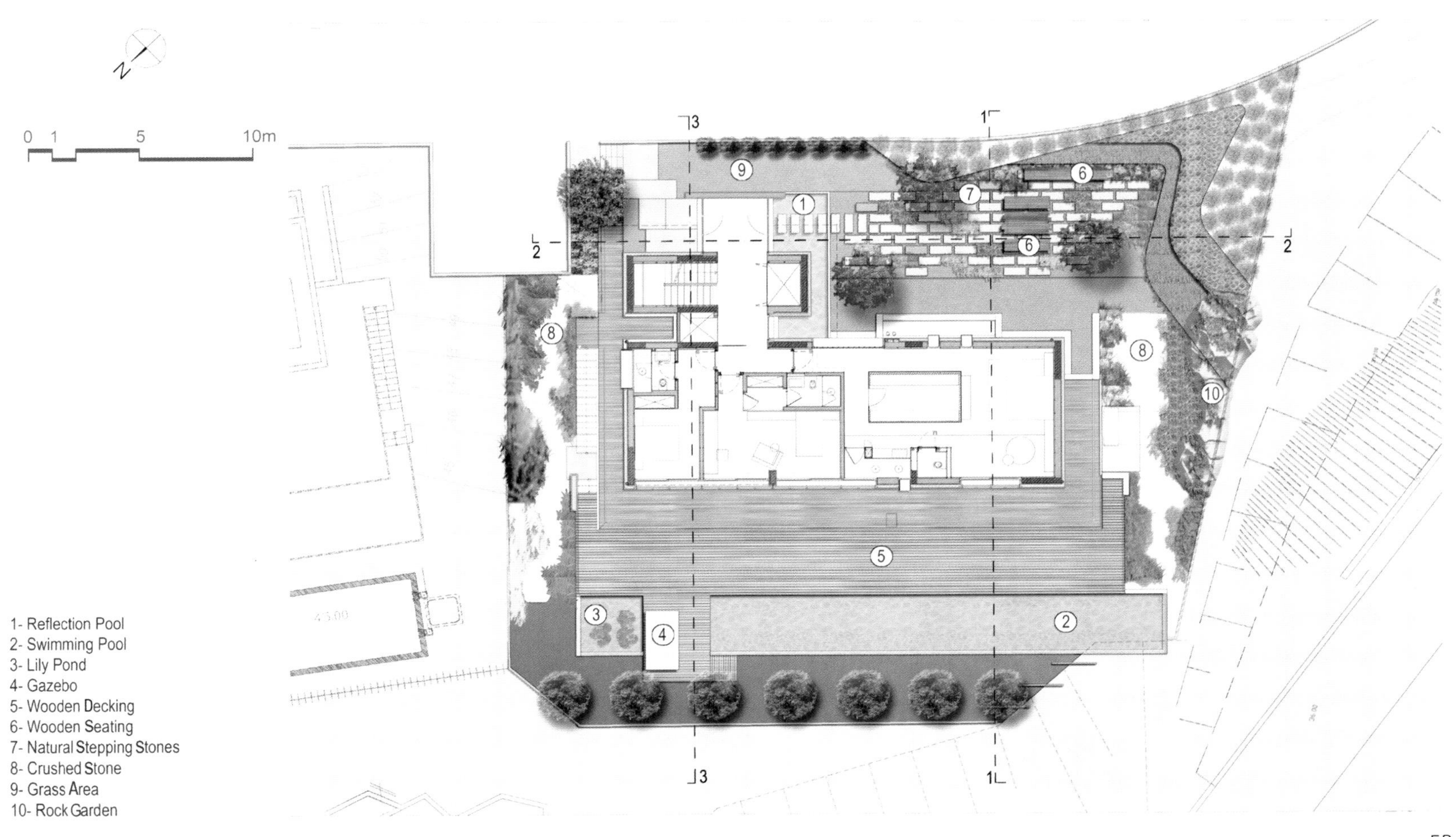
N
0 1 5 10m
1- Reflection Pool
2- Swimming Pool
3- Lily Pond
4- Gazebo
5- Wooden Decking
6- Wooden Seating
7- Natural Stepping Stones
8- Crushed Stone
9- Grass Area
10- Rock Garden

Turtle Rock

Landscape Architect
Studio H Landscape Architecture, Inc.

Site Elements
Pool, Spa, Fireplace, Fire Pit, Barbecue, Outdoor Bar, Balcony, Patio Cover, Gabion Walls, Zen Garden, Succulent Planting, Ipe Wood Deck

Location
Irvine, California, USA

Photographer
Studio H Landscape Architecture

This project is located in the Turtle Ridge community near UCI in Irvine. With an existing pool, the goal was to create a modern inviting space with multiple entertaining areas. The unique areas include a fire pit and lounging area at one end of the pool with Tuscan styled planting, at the other end is a modern pass-through fireplace with fire balls and a custom water fountain spilling into the pool. The fireplace wraps around to include a barbecue and bar area with an Ipe wood deck seating lounge. Just off the bar area is a Zen Garden with custom water features spilling from the gabion walls, a pebble contemplation garden, and bamboo planting. As you pass through the Zen Garden you enter the Tuscan themed entertaining space with corten steel planters, citrus trees, and decomposed granite.

A few of the materials included in this project are as follows: Limestone paving, concrete coping, gabion walls, limestone veneer, concrete countertops, custom stainless steel scuppers, 25.4mm glass tile veneer, large wood beams, glass rails, Ipe wood, rolled pebble and succulent planting.

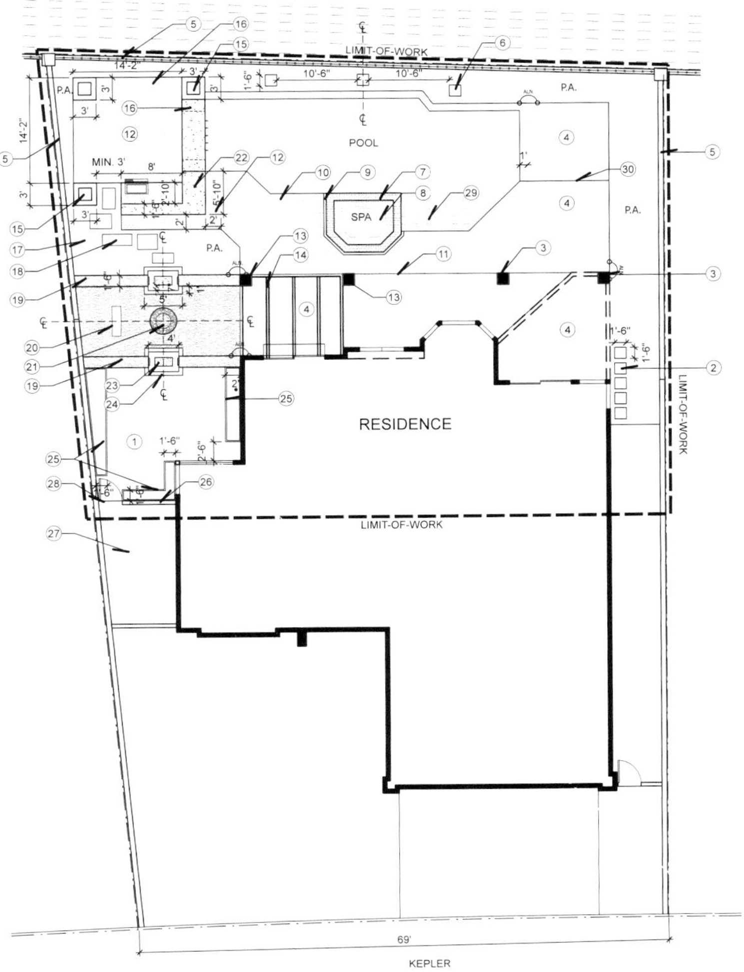

CONSTRUCTION CALLOUTS:

1. DECOMPOSED GRAVEL PAVING - BUFF COLOR
2. 18" X 18" COLORED CONCRETE STEPPING PAD
3. ADD STONE VENEER TO EXISTING 18" STUCCO COLUMN. 24" HIGH
4. 24" SQ. LIMESTONE PAVERS ON CONC. BASE, SEALED.
5. REMOVE EXISTING ALL W/ FENCE AND REPLACE W/ 5' TALL CMU BLOCK WALL W/ STUCCO FINISH AND BLOCK CAP TO MATCH EXISTING PL WALL
6. 18" X 18" CONCRETE POTTERY PEDESTAL
7. 18" RAISED SPA BOND BEAM WITH 1/2" GLASS TILE AND INFINITY EDGE. TILE TO EXTEND 6" FROM TOP OF BOND BEAM ON INTERIOR.
8. EXISTING SPA- RAISE 18" AND MOVE EXISTING JETS AND LIGHTS, RE-PLASTER WITH PEBBLETECH OR SIMILAR TO MATCH POOL.
9. SPACING PER CONTRACTOR
10. 12" CAST-IN-PLACE COLORED CONCRETE COPING WITH LIGHT SAND BLAST FINISH - STRAIGHT EDGE
11. PROPOSED BALCONY W/ GLASS RAILING
12. 4" WIDE X 1" THICK SCR IPE WOOD DECKING W/ PENIFIN OIL STAIN FINISH. DARK NATURAL WOOD COLOR. USE HIDDEN BRACKETS
13. PROPOSED 18" COLUMN TO MATCH EXISTING. ADD STONE VENEER TO 24" HIGH.
14. PROPOSED OPEN PATIO COVER .
15. 18" TALL PLANTER
16. CMU BLOCK FIREPLACE W/ 12"X24" LIMESTONE VENEER
17. DECORATIVE PEA GRAVEL OR DECOMPOSED GRANITE, T.B.D.
18. SCR IPE WOOD STEPPING PADS W/ PENIFIN DARK STAIN/SEALER. USE INVISIBLE CLIPS FOR INSTALLATION. MIN. 8" SPACING BETWEEN PADS. APPROXIMATE SIZE SHOWN, RESIZE AS NEEDED FOR THE SPACE.
19. 18" WIDE X 24" TALL GABION WALL W/ NATURAL STONE AND BENT IRON REBAR CAGE.
20. PRECAST DECORATIVE BENCH
21. BUBBLING URN WATER FEATURE
22. BAR/BARBECUE AREA W/ CIP CONCRETE COUNTERS AND GLASS/LIGHTING RACK.
23. COLORED CONCRETE STEP PAD W/ LIGHT SANDBLAST FINISH - LIGHT BEIGE COLOR FINISH
24. CIP CONCRETE BASIN W/ WATERPROOFING AND 1/2" GLASS TILE TO MATCH SPA.
25. 18" RAISED STEEL VEGETABLE PLANTERS - CORETEN STEEL ALLOWED TO RUST
26. EXISTING CMU BLOCK WALL W/ STUCCO FINISH
27. PEA GRAVEL AROUND POOL EQUIPMENT.
28. EXISTING WOOD GATE
29. EXISTING POOL STEPS TO MATCH POOL FINISH.
30. LIMESTONE STEP W/ LIMESTONE VENEER ON RISER. MITER EDGES

LEGEND:

ALN	ALIGN
CLR	CLEAR
℄	CENTERLINE
EQ.	EQUAL
F.O.C.	FACE OF CURB
F.O.W.	FACE OF WALL
MIN.	MINIMUM
MAX.	MAXIMUM
P.A.	PLANTING AREA
R=	RADIUS EQUALS
TYP.	TYPICAL
(18)	CONSTRUCTION CALLOUT
•	CAULKED EXPANSION JOINT (TO MATCH PAVING)

FIELD VERIFY ALL UTILITY LOCATIONS PRIOR to CONSTRUCTION

GENERAL NOTES:

- ALL DRAINPIPES MUST DRAIN AT A MINIMUM OF 1/2%. DRAINLINES ON PLAN ALL DRAIN AT 2%.
- ALL PLANTER BEDS AND PAVED AREAS TO DRAIN AT MINIMUM OF 2% OR 1/4" PER FOOT.
- ALL PLANT BEDS AND PAVING TO SLOPE AND DRAIN AWAY FROM HOUSE.
- ALL GRADES OF PLANT BEDS MUST BE HELD A MINIMUM OF 6' BELOW THE TOP OF ADJACENT PLANTER OR RETAINING WALL.
  ALL GRADES OF PLANT BEDS MUST BE HELD A MINIMUM OF 6' BELOW METAL
- HOUSE SCREED. ALL HARDSCAE MUST BE HELD 4" BELOW METAL HOUSE SCREED.
  DON'T RETAIN DIRT AGAINST ANY PERIMETER WALL UNLESS A SUBWALL IS
- PROVIDED.

LEGEND:

1. MITER LIMESTONE CORNERS
2. MODERN 3/4" SPIGOT- T.B.D
3. LIMESTONE VENEER - SIZE VARIES, BUTT JOINT
4. GLASS TILE ON POOL - TBD
5. GLASS FIRE ROCK
6. 4" IPE WOOD VENEER ON P-I-P CONCRETE PLANTER
7. P-I-P CONCRETE
8. EXISTING BLOCK WALL WITH FENCE
9. P-I-P CONCRETE COUNTER, POLISHED AND SEALED
10. BOWL PLANTER W/ DRIP IRRIGATION AND DRAINAGE
11. 12"X24" LIMESTONE VENEER , BUTT JOINT, NO GROUT
12. 4" CMU BLOCK WALL, REBAR PER CONTRACTOR
13. 1" GLASS TILE BY OCEANSIDE
14. STAINLESS STEEL UTILITY CABINETS

NOTE:

1. FINISH COLORS TO BE SELECTED.
2. REINFORCED AS REQUIRED
3. STONE TO BE MITERED AT CORNERS

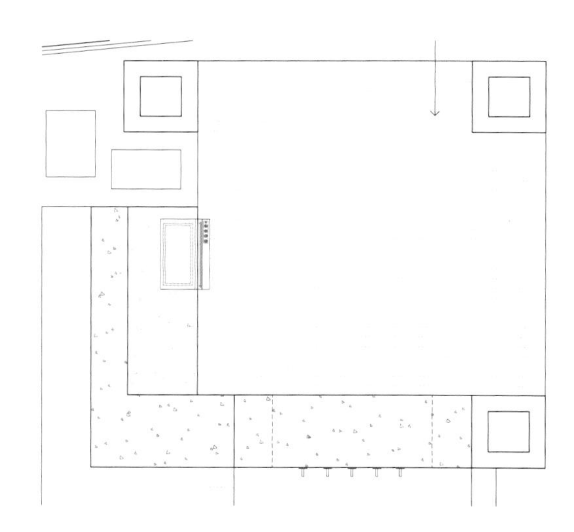

PLAN VIEW

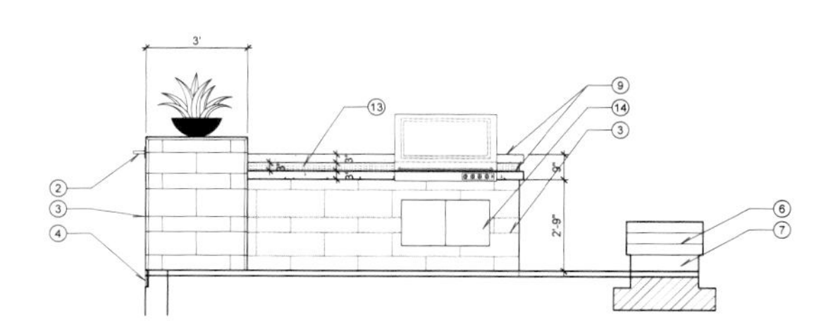

ELEVATION C

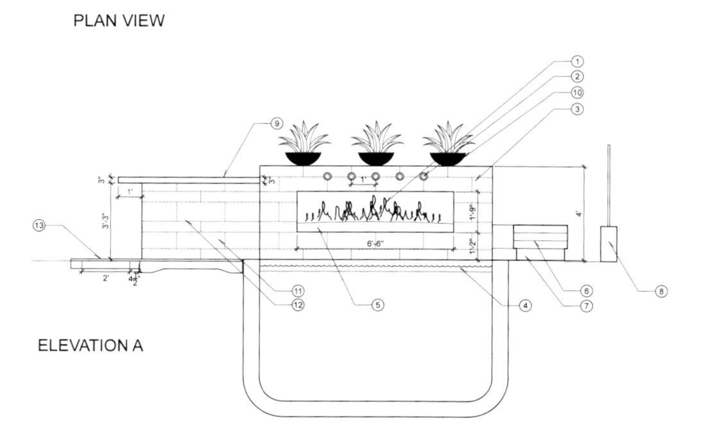

ELEVATION A

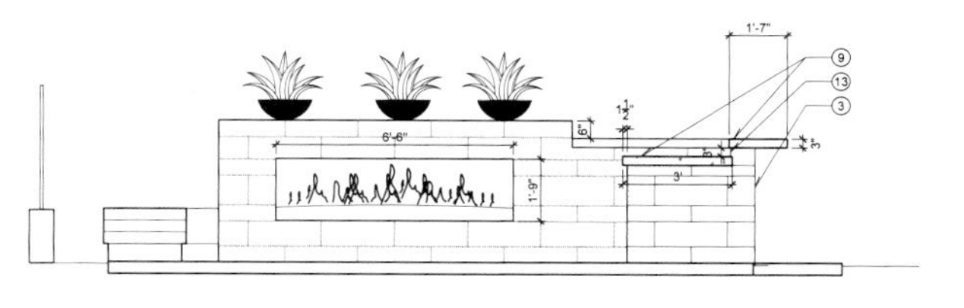

ELEVATION B

(A) FIREPLACE ELEVATION

Scale: 1/2" = 1' - 0"

LEGEND:

1. 2" STAINLESS STEEL FIREBOX OR TRIM
2. MODERN 3/4" SPIGOT
3. 12" X 24" SANDSTONE PAVER VENEER
4. GLASS TILE ON POOL - TBD
5. GLASS FIRE ROCK
6. 4" IPE WOOD VENEER ON P-I-P CONCRETE PLANTER
7. P-I-P CONCRETE
8. EXISTING BLOCK WALL WITH FENCE
9. P-I-P CONCRETE COUNTER, POLISHED AND SEALED
10. BOWL PLANTER W/ DRIP IRRIGATION AND DRAINAGE

NOTE:

1. FINISH COLORS TO BE SELECTED.
2. REINFORCED AS REQUIRED
3. STONE TO BE MITERED AT CORNERS

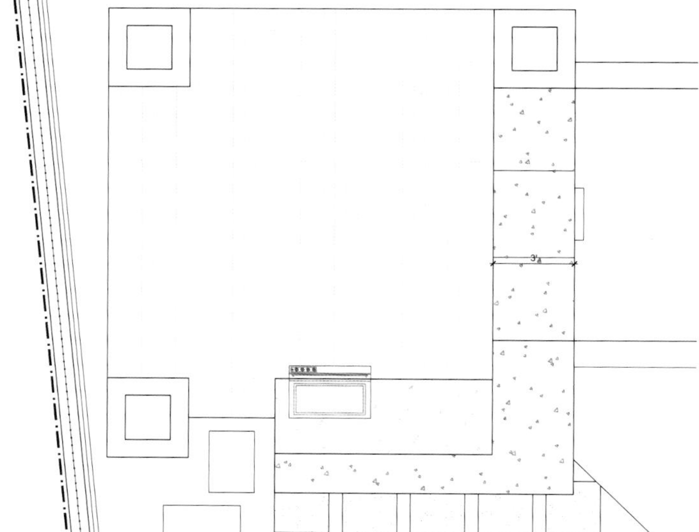

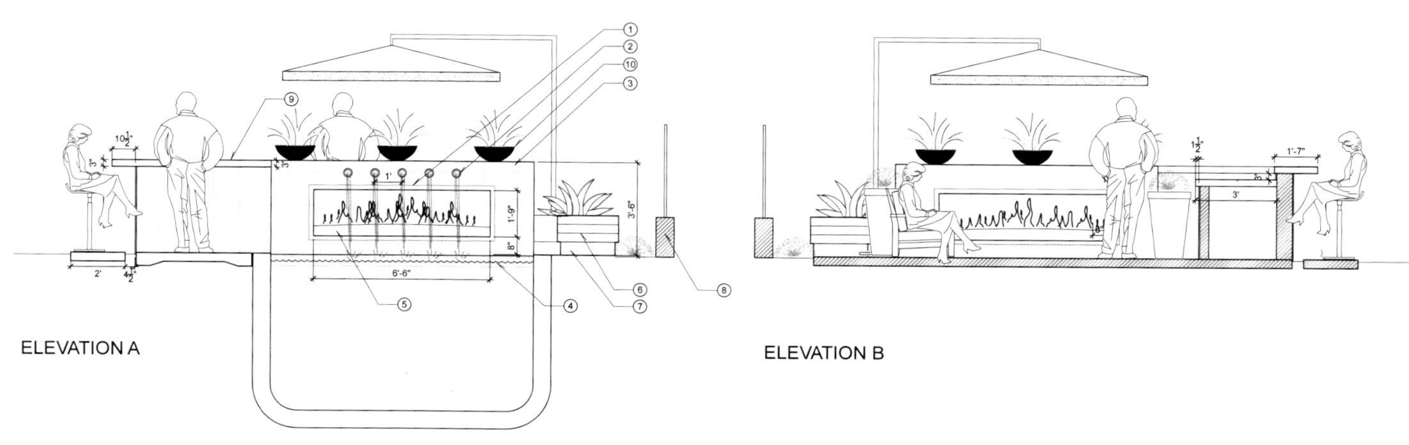

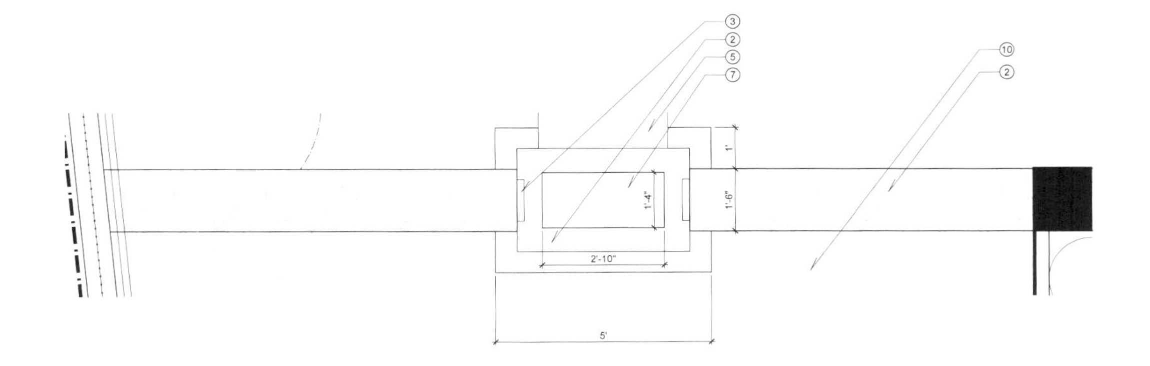

PLAN VIEW

SECTION A

LEGEND:

1. #4 REBAR WITH STEEL TIES
2. 4" - 6" ROLLED RIVERROCK - BUFF COLORED
3. STAINLESS STEEL 8" WIDE SHEER DESCENT
4. 3/4" COPPER OVERFLOW/DRAIN W/ THREADED BOTTOM FOR DRAINAGE.
5. IPE WOOD BRIDGE W/ LED LIGHTING
6. 4" WIDE WATER PROOF CONCRETE BASIN - 2500 PSI @ 28 DAYS - SIZE PER ENGINEER. REBAR PER STRUCTURAL ENGINEER. SIDEWALLS @ 4" WIDE
7. COLORED CONCRETE W/ LIGHT SANDBLAST FINISH STEPPING PAD AND BORDER
8. UNDERWATER LIGHT FIXTURE
9. #4 REBAR, RUSTED AND SEALED
10. WHITE ROLLED PEA GRAVEL
11. 2"- SQUARE STEEL CAGE, RUSTED AND SEALED TIED TO #4 REBAR
12. 18" COLUMN W/ STUCCO FINISH TO MATCH BUILDING

NOTE:

1. FOUNTAIN CONTRACTOR TO PROVIDE SHOP DRAWINGS TO OWNER & LANDSCAPE ARCHITECT PRIOR TO CONSTRUCTION.
2. BASIN TO BE POURED AGAINST UNDISTURBED SOIL OR RECOMPACTED SOIL PER STRUCTURAL SOILS REPORT.

Dural

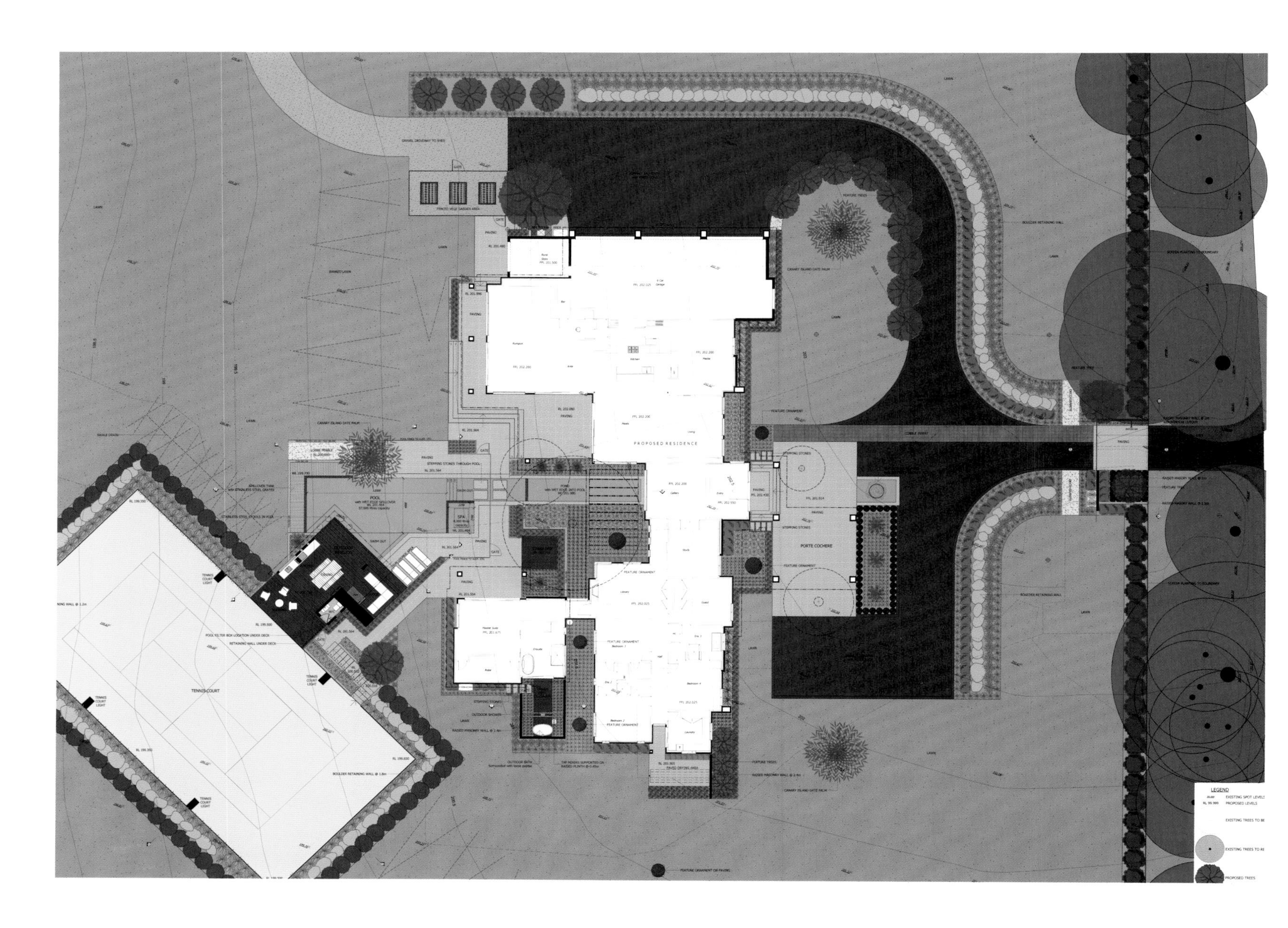

Landscape Architect
Dean Herald

Landscape Construction
Rolling Stone Landscapes

Location
Dural, New South Wales, Australia

Photographer
Danny Kildare

Located on a 20,234.28m^2 property in Dural the landscape for this project was a blank canvas with the owners building their new dream home. The client's brief was to have a strong indoor-outdoor connection along with making the most of the outdoors while entertaining family and friends.

Working with the house designer it was ensured large glass viewports were created forming strong site lines to the landscape from a number of locations within the home. This was an essential platform to allow a strong indoor-outdoor connection be achieved as per the client brief.

The pool was positioned to capture your attention as you entered the home. A shallow pond, with floating steppers, flows into the pool. Accommodating a spa the pool then stretches a further 14.5m, before reaching the infinity edge.

Adjoining the pool is the fully equipped pavilion. Featuring lounge, dining and kitchen areas it is designed to provide the perfect space for interaction with family and friends. Being designed at a 45 degree angle provides for a great line of sight from the kitchen area. The pavilion backs onto a tennis court providing a perfect elevated position to view a friendly tennis match.

Plants used were selected to give a relaxed feel with undertones of luxury. An advanced Canary Island date palm was positioned next to the pool to provide midday shade and give a horticultural balance to the pavilion structure.

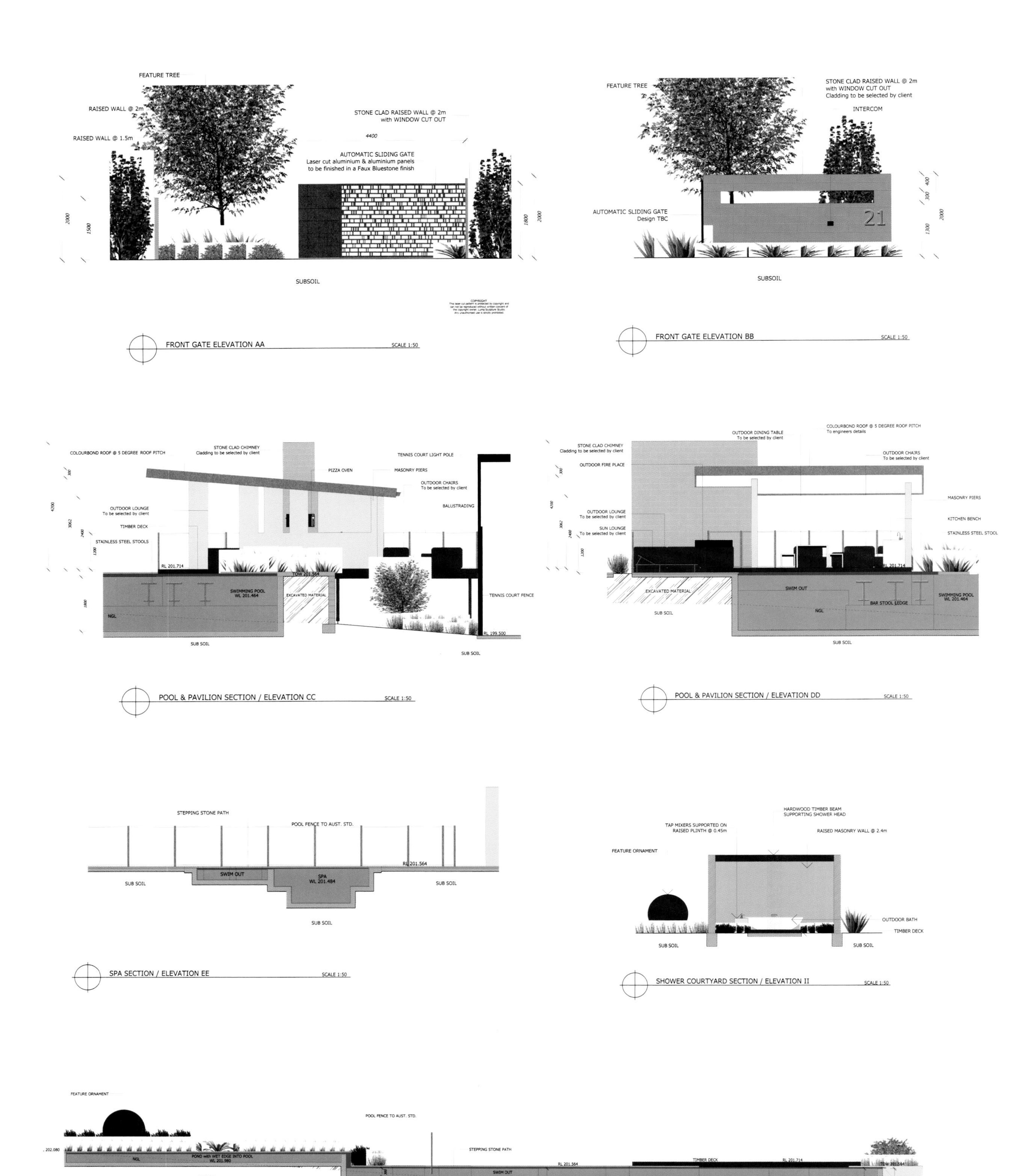
FEATURE TREE
RAISED WALL @ 2m
RAISED WALL @ 1.5m
STONE CLAD RAISED WALL @ 2m with WINDOW CUT OUT
AUTOMATIC SLIDING GATE
Laser cut aluminium & aluminium panels to be finished in a Faux Bluestone finish
SUBSOIL
FRONT GATE ELEVATION AA
SCALE 1:50
Cladding to be selected by client
INTERCOM
AUTOMATIC SLIDING GATE
Design TBC
FRONT GATE ELEVATION BB
COLOURBOND ROOF @ 5 DEGREE ROOF PITCH
STONE CLAD CHIMNEY
PIZZA OVEN
TENNIS COURT LIGHT POLE
MASONRY PIERS
OUTDOOR CHAIRS
To be selected by client
OUTDOOR LOUNGE
TIMBER DECK
STAINLESS STEEL STOOLS
BALUSTRADING
SWIMMING POOL
WL 201.464
EXCAVATED MATERIAL
TENNIS COURT FENCE
SUB SOIL
POOL & PAVILION SECTION / ELEVATION CC
OUTDOOR DINING TABLE
To engineers details
OUTDOOR FIRE PLACE
SUN LOUNGE
KITCHEN BENCH
SWIM OUT
BAR STOOL LEDGE
POOL & PAVILION SECTION / ELEVATION DD
STEPPING STONE PATH
POOL FENCE TO AUST. STD.
SPA
WL 201.484
SPA SECTION / ELEVATION EE
HARDWOOD TIMBER BEAM SUPPORTING SHOWER HEAD
TAP MIXERS SUPPORTED ON RAISED PLINTH @ 0.45m
RAISED MASONRY WALL @ 2.4m
FEATURE ORNAMENT
OUTDOOR BATH
SHOWER COURTYARD SECTION / ELEVATION II
POND with WET EDGE INTO POOL
WL 201.980
RENDERED RAISED WALL
To match house colour
SPILLOVER TANK
With safety grate & pebbles
POOL SECTION / ELEVATION FF

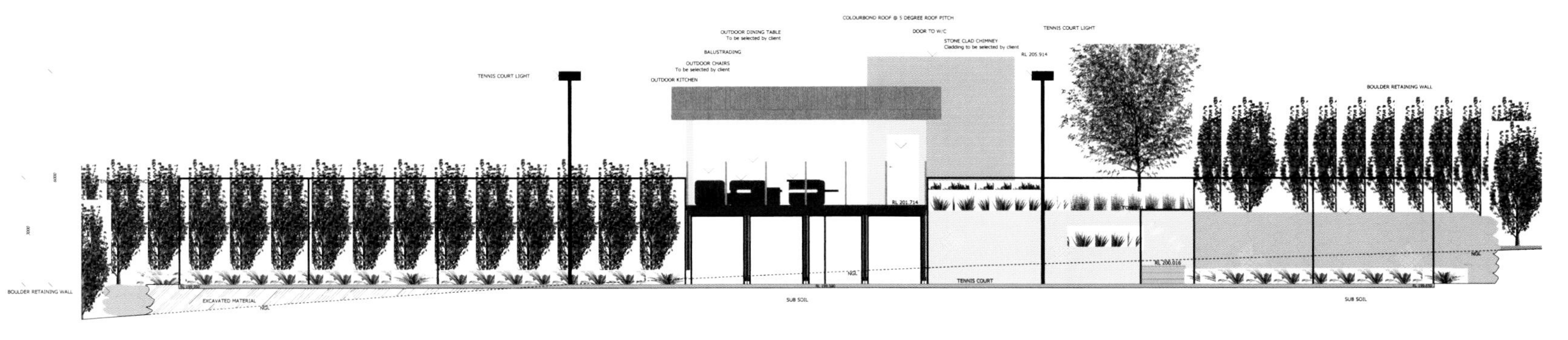

TENNIS COURT & PAVILION SECTION / ELEVATION GG SCALE 1:50

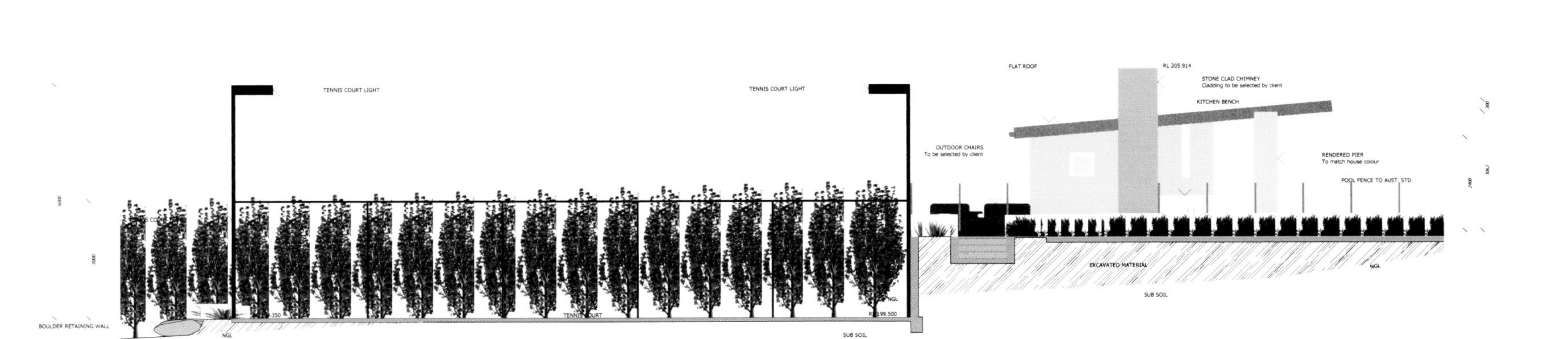

TENNIS COURT & PAVILION SECTION / ELEVATION HH SCALE 1:50

Panoramic Terrace

Landscape Architect
Steve Taylor

Firm
COS Design

Photographer
Tim Turner

After many years of building a very successful business and living very humbly in their original AV Jennings 19m^2 home, this Melbourne family decided it was time to design a pool and surrounding space that epitomised a resort style environment with a lap lane, spa and large family play space. Levels were daunting with over 19m of fall from front to rear on this property and the house design, in particular the facade/style had been back and forth for eighteen months between architect and client. In the end COS Design was commissioned to design the pool and garden to fulfil the client's ultimate dream outdoor space.

With large expansive spaces throughout the rest of the property, the clients wanted to create a private, tranquil adult space to relax, read a book or the paper with a morning coffee. The small courtyard runs directly off the internal meals area and is the focal point from this highly used internal vista. The brief was to create a functional, visually stunning, private courtyard space with some form of moving water to create a calming ambience.

Front Yard Elevation
SCALE 1:100

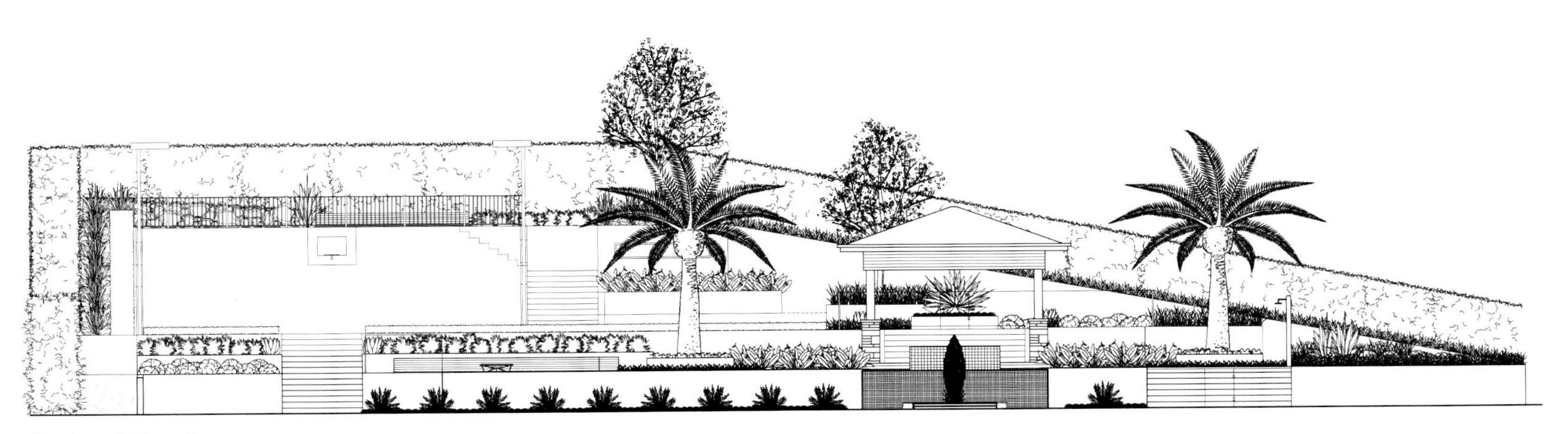

Backyard Elevation
SCALE 1:100

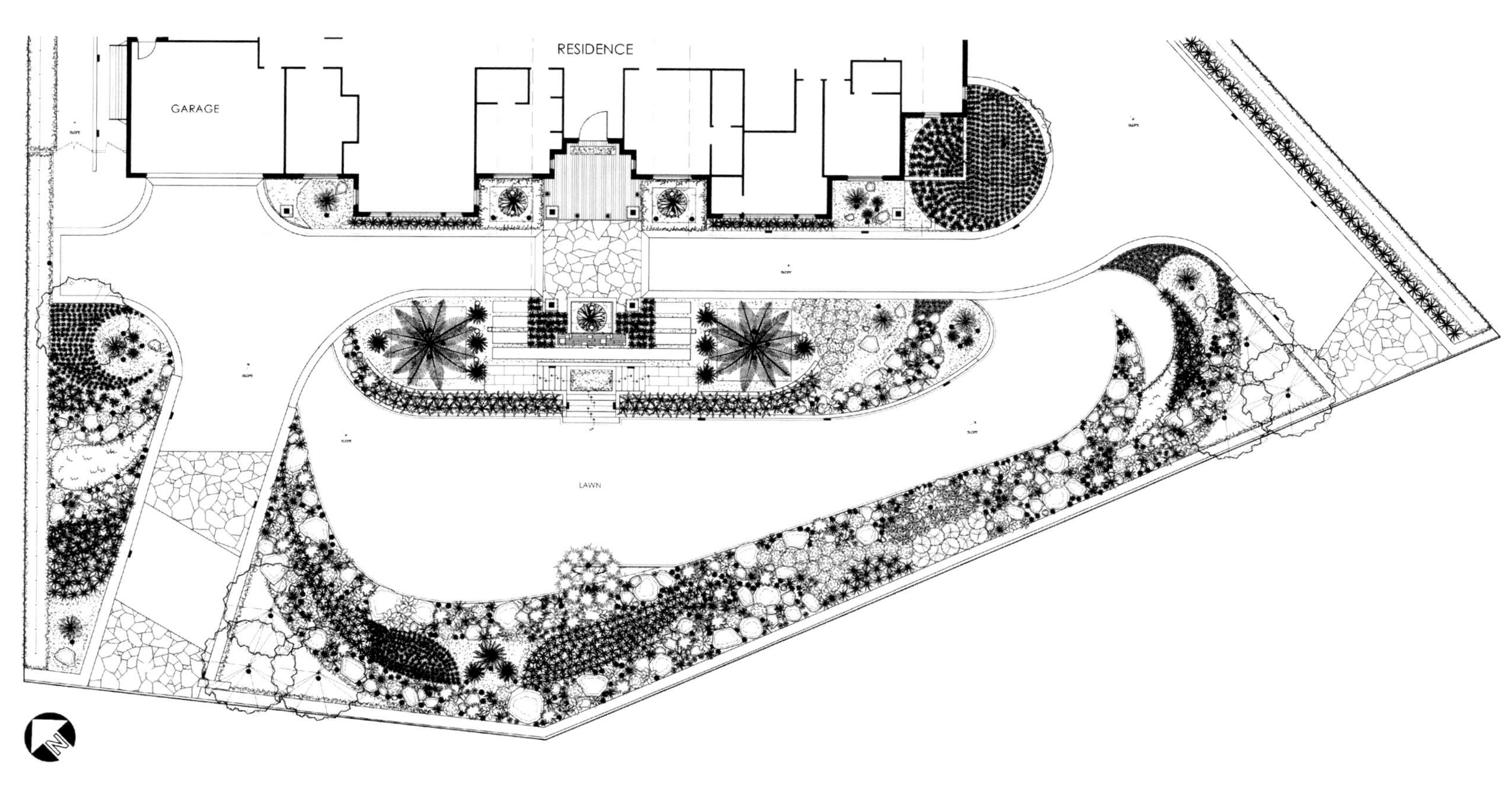
RESIDENCE
GARAGE
LAWN

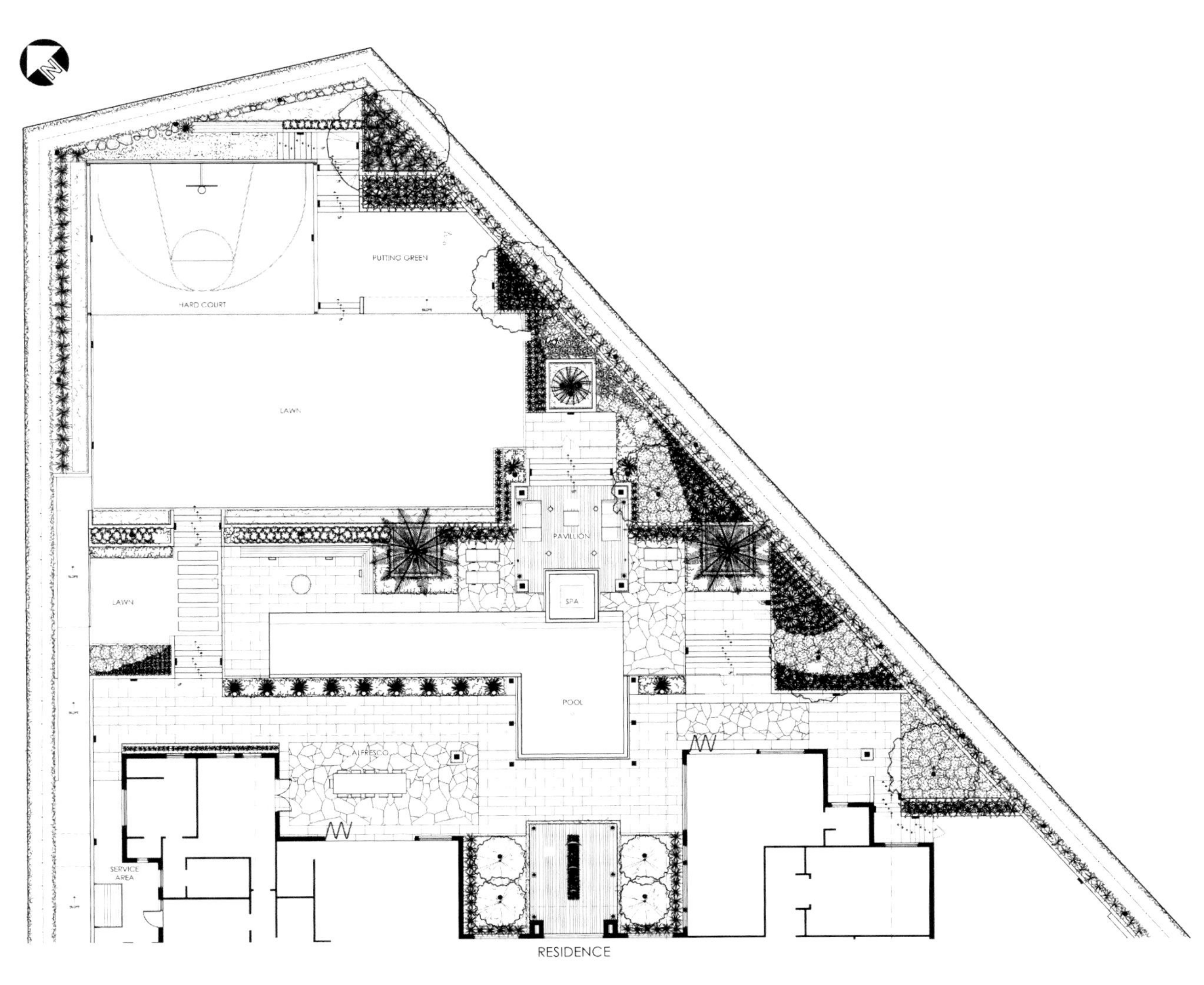
PUTTING GREEN
HARD COURT
LAWN
PAVILION
SPA
LAWN
POOL
ALFRESCO
SERVICE AREA
RESIDENCE

Knuppel Residence

Landscape Architect
Studio H Landscape Architecture, Inc.

Location
Newport Coast, California, USA

Photographer
Isabella Li

This Mediterranean styled home is located in Newport Coast, California. The clients loved their home but found they never used the rear yard and were looking for a design that would draw them into the space. The central element of the project is the raised pool/spa/fountain combination which is a focal from both ends of the yard as well as from the interior views. The outdoor entertaining space off the living room features a large gas fireplace, rustic wood patio cover with concrete columns and a large built-in barbecue. There is a secondary patio area off the master bedroom with a secondary rustic shade structure.

Elements are Gas Fireplace, barbecue area, raised pool/spa combination with fountain, dining area and seating areas, rustic wood patio covers with concrete columns, terra cotta pottery, low water-use plant material, LED lighting, and travertine paving.

LEGEND:

ALN	ALIGN
CLR	CLEAR
℄	CENTERLINE
EQ.	EQUAL
F.O.C.	FACE OF CURB
F.O.W.	FACE OF WALL
MIN.	MINIMUM
MAX.	MAXIMUM
P.A.	PLANTING AREA
R=	RADIUS EQUALS
TYP.	TYPICAL
(18)	CONSTRUCTION CALLOUT
•	CAULKED EXPANSION JOINT (TO MATCH PAVING)

FIELD VERIFY ALL UTILITY LOCATIONS PRIOR to CONSTRUCTION

CONSTRUCTION CALLOUTS:

1. RANDOM FLAGSTONE PAVING ON CONCRETE BASE - BUFF COLOR
2. CMU BLOCK BARBECUE W/ STUCCO SMOOTH STUCCO FINISH
3. WOOD PATIO COVER W/ STAIN TO MATCH TRIM ON RESIDENCE
4. EXISTING BLOCK PROPERTY LINE WALL
5. COLORED CONCRETE W/ LIGHT SANDBLAST FINISH AND SAW-CUT SCORE JOINTS. SEE FINISH SCHEDULE FOR FINISH COLOR
6. PRECAST CONCRETE FIREPLACE BY...
7. SURLOCK STEEL LANDSCAPE EDGING
8. 12" COLORED CONCRETE COPING W/ BULLNOSE, LIGHT SANDBLAST FINISH. SEE FINISH SCHEDULE
9. 12" X 24" COLORED CONCRETE STEPPING PAD. SEE FINISH SCHEDULE.
10. POOL FOUNTAIN BY...
11. MASTER BEDROOM WOOD PATIO COVER, STAIN TO MATCH TRIM ON RESIDENCE. SEE FINISH SCHEDULE

FINISH SCHEDULE:

CALLOUT	ITEM	MANUFACTURER/SUPPLIER	COLOR	FINISH	COMMENTS
(1)	CONCRETE PAVING	L.M. SCOFIELD CO. (800) 800-9900	TBD	MEDIUM RELEASE FINISH	WITH SAWCUT JOINTS AS SHOWN
(1)	4" CERAMIC TILE	TO BE SELECTED BY OWNER	TO MATCH ARCHITECTURE		
(1)(1)(1)	NATURAL STONE PAVING/STEPS	TO BE SELECTED BY OWNER	TO BE SELECTED BY OWNER	HONED - SLIP RESISTANCE - MIN COF +.70)	
(1)(1)	CONCRETE CAP / HEARTH	L.M. SCOFIELD CO. (800) 800-9900	TBD	MEDIUM RELEASE FINISH	WITH SAWCUT JOINTS AS SHOWN
(1)	POOL TILE	TO MATCH FIREPLACE/BBQ	TO MATCH FIREPLACE / BBQ	CERAMIC	
(1)	PLASTER WALLS	TO BE SELECTED BY OWNER	TO MATCH ARCHITECTURE	TO MATCH ARCHITECTURE	
(1)	WOOD PERGOLA STAIN	TO BE SELECTED BY OWNER	TO MATCH ARCHITECTURE	TO MATCH ARCHITECTURE	
(1)	WOOD GATE	TO BE SELECTED BY OWNER	TO BE SELECTED	TO BE SELECTED	

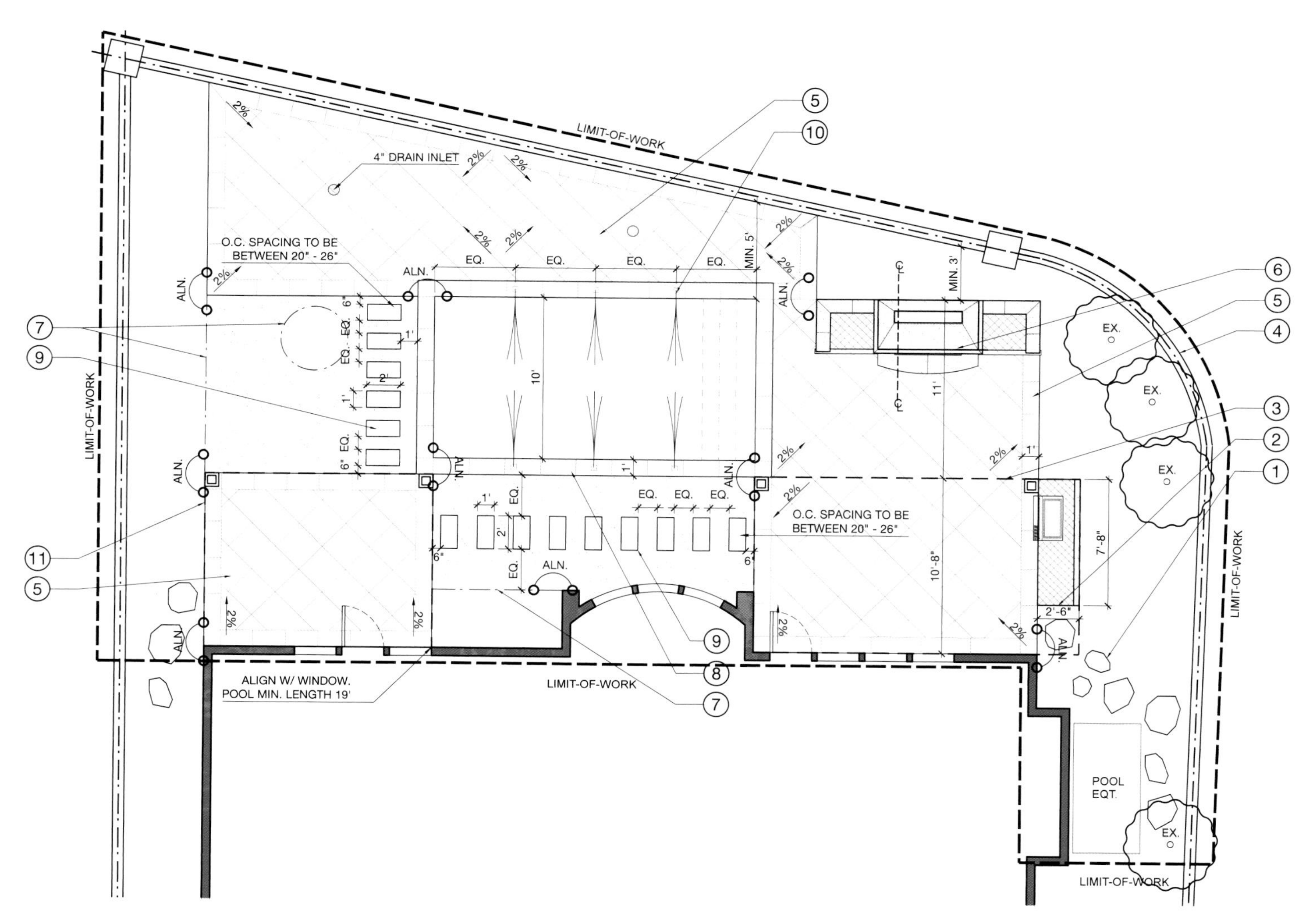

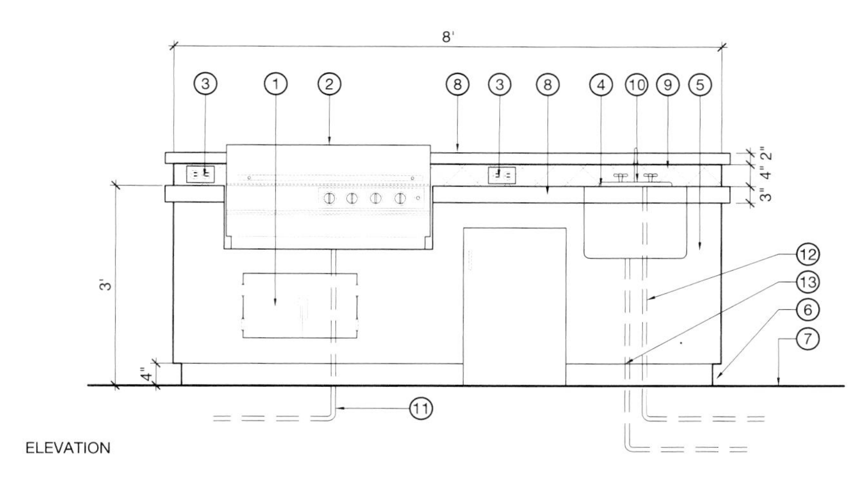

ELEVATION

3'
6"
4" 2"
1"

SECTION

LEGEND:

1. STAINLESS STEEL ACCESS DOORS
2. DROP IN GAS BBQ MODEL. TO BE SELECTED BY OWNER. INSTALL PER MANUFACTURER'S SPECIFICATIONS
3. ELECTRICAL OUTLET
4. COUNTER SURFACE - TO MATCH KITCHEN COUNTER
5. C.M.U. BASE with SMOOTH PLASTER.
6. 4" KICKPLATE
7. FINISH SURFACE
8. COLORED CONCRETE CAP, LIGHT SANDBLAST FINISH.
9. 4" DECORATIVE TILE - TO BE SELECTED
10. STAINLESS STEEL SINK - MODEL TO BE SELECTED BY OWNER
11. GAS SERVICE LINE BY PLUMING CONTRACTOR
12. POTABLE WATER LINE BY PLUMING CONTRACTOR
13. CONNECT INTO EXISTING SYSTEM PER PLUMBING CONTRACTOR
14. CERAMIC TILE - TO BE SELECTED BY OWNER

A C.M.U. BARBECUE WITH STUCCO FINISH

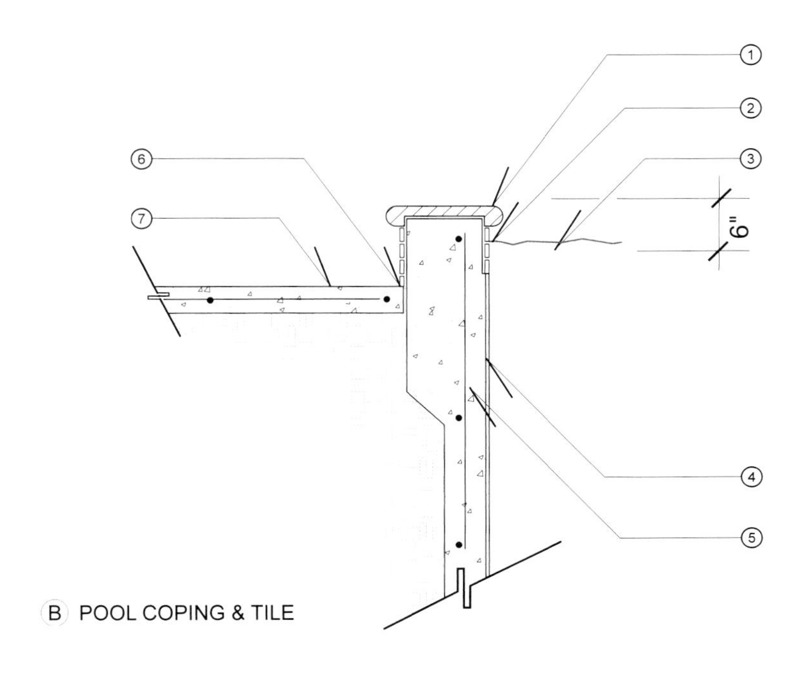

LEGEND:

1. 12"X12"X 3" BULLNOSE POOL COPING
2. 3 COURCES 2" x 2" CERAMIC TILE SET IN MORTAR BED
3. WATER LEVEL
4. MIN 1/2" THICK POOL PLASTER COLOR: WHITE
5. POOL/SPA BOND BEAM PER STRUCTURAL ENGINEER
6. ELASTOMERIC EXPANSION JOINT CONTINUOUS BEHIND COPING COLOR TO MATCH POOL DECK PAVING
7. CONCRETE POOL DECK

B POOL COPING & TILE

LEGEND:

1. #3 BAR @ 24" O.C. - CENTER IN SLAB
2. CONCRETE PAVING W/ SAWCUTS PER PLAN
3. 12"x 24" PAVER BANDING OVER THIN SET
4. FINISH GRADE
5. COMPACTED SUBGRADE

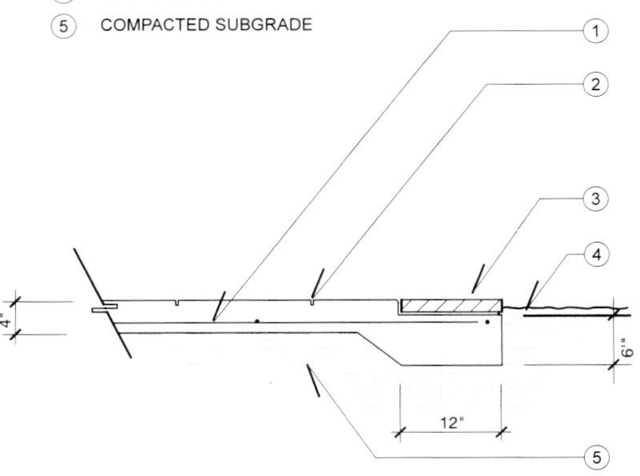

C CONC. PAVING W/ PAVER BANDING 1"=1'0"

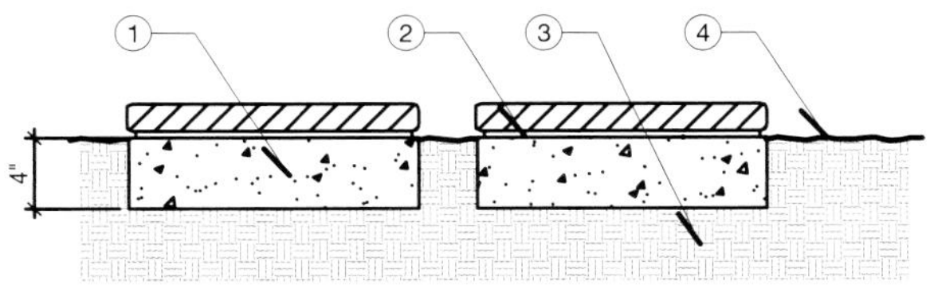

SECTION

LEGEND:

1. CONCRETE BASE
2. STONE OVER THINSET
3. COMPACTED SUBGRADE
4. FINISH GRADE

D STONE STEPPING PADS Scale: 1 1/2" = 1' - 0"

LEGEND:

1. 4 x 4 TILE T.B.S. FRONT AND SIDES ONLY
2. STUCCO FINISH
3. 2' WD. BULLNOSE TILE
4. WOOD UTILITY DOORS W/ DARK STAIN FINISH
5. METAL FLASHING W/ LATHE POP-OUT W/ STUCCO FINISH @ FIREPLACE OPENING
6. 6" HT. BACK SPLASH W/ STUCCO FINISH
7. CONCRETE FOOTING - 2500 PSI @ 28 DAYS - SIZE PER ENGINEER
8. 6x8x16 C.M.U. BLOCK
9. REBAR HORIZ. SIZE & LOCATION PER ENG.
10. COMPACTED SUBGRADE PER SOILS REPORT.
11. VERT. REBAR @BEND INTO FOOTING AS SHOWN INTO FOOTING- SIZE PER STRUCTURAL ENG.
12. POURED IN PLACE CONCRETE
13. SCREEN FILTER TO PREVENT ASH DRIFT
14. NON-COMBUSTABLE LINTEL - FLAT IRON
15. 4 x 2 x 8 FIREBRICK ALL VISIBLE INTERIOR SIDES HERRINGBONE PATTERN
16. FINISH SURFACE
17. TERRA COTTA FLUE
18. CONCRETE WALK. CONTINUOUS E.J. @ BASE OF FIREPLACE.
19. WOOD POP-OUT W/ STUCCO FINISH

NOTES:

1. ALL STUCCO TO BE SMOOTH FINISH, COLOR TO MATCH ARCHITECTURE
2. MAX. HEIGHT NOT TO EXCEED 6'
3. P.L. SETBACK TO BE MIN. 3'

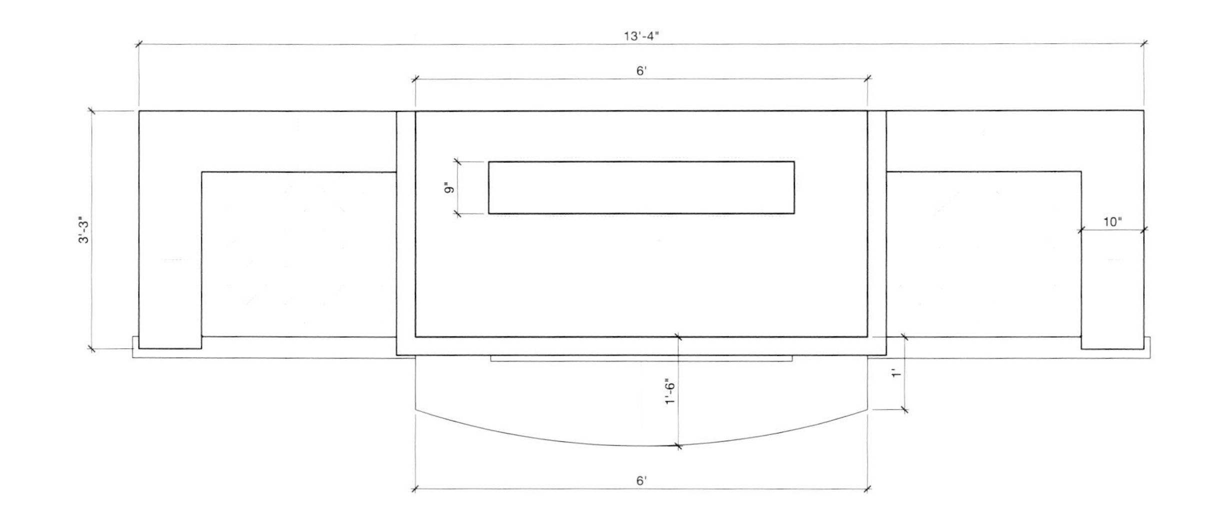

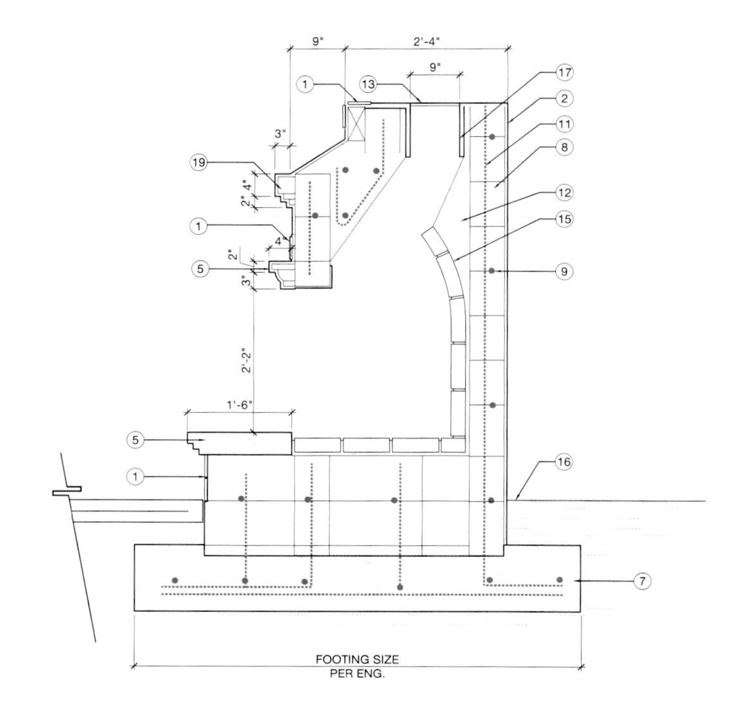

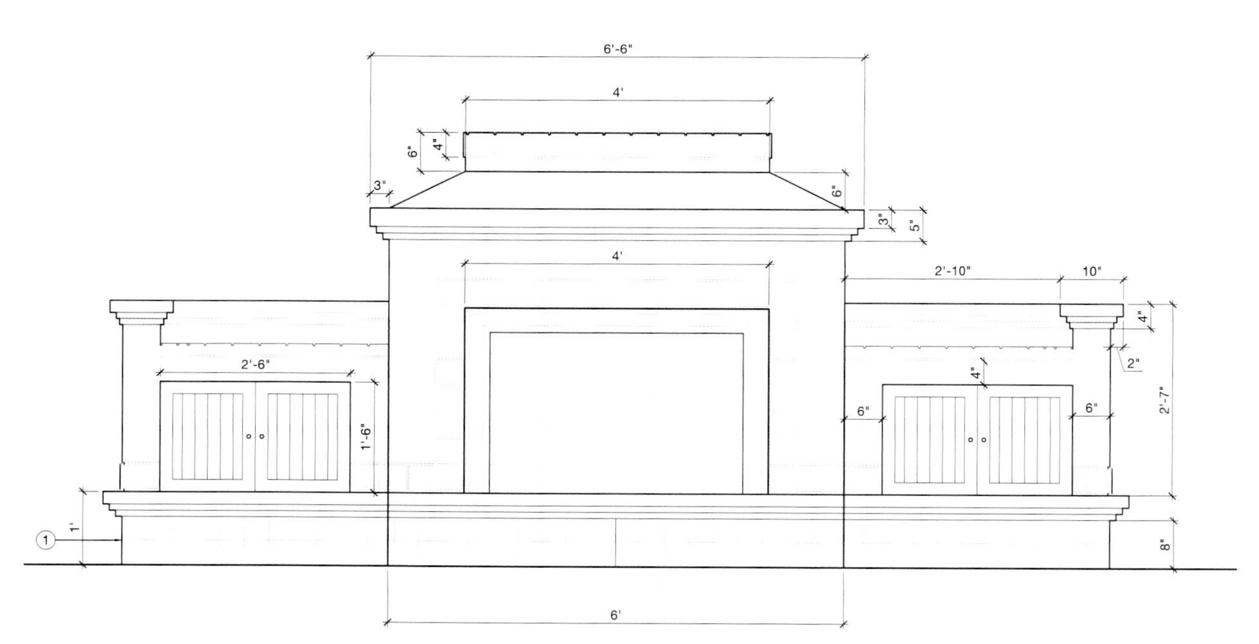

OUTDOOR FIREPLACE W/ STUCCO FINISH & TILE

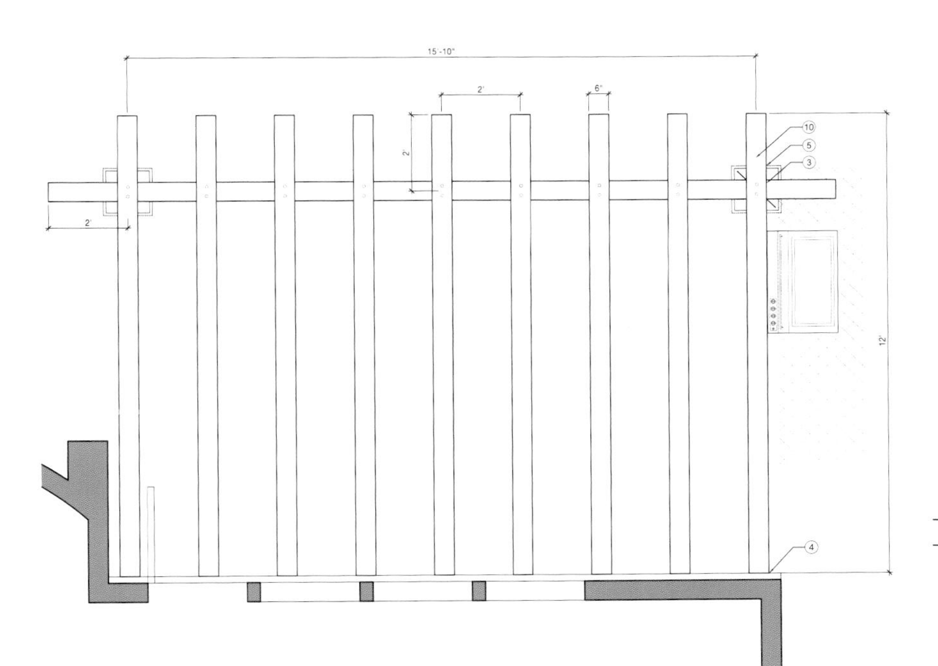

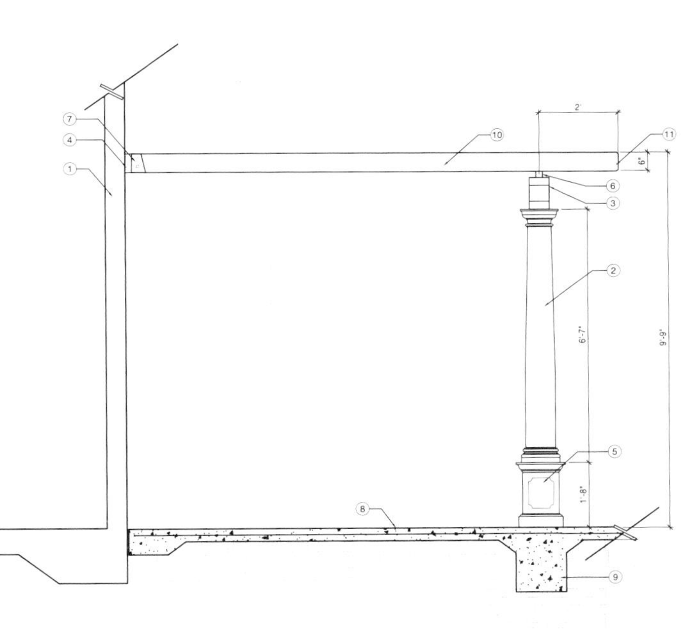

LEGEND:

1. HOUSE WALL
2. HADDONSTONE COLUMN
   M800A/F PLAIN SHAFT 74" 3/8" INCLUDING BASE AND CAPITAL
3. 6 x 10 STANDARD REDWOOD BEAM
4. 2 x 4 LEDGER
   ATTACH TO HOUSE W/ LAG BOLTS 16" O.C.
5. HADDONSTONE PLINTH
   B400 ELIZABETHAN PLINTH 20"
6. 2 STEAL SPACER- SIMPSON
7. SIMPSON JOIST HANGER
8. FINISH SURFACE
9. CONCRETE FOOTING - 2500 PSI
   @ 28 DAYS - SIZE PER CONTRACTOR
10. 6 x 6 STANDARD REDWOOD LATTICE
11. CAMPHORED EDGE ON LATTICE

NOTE:

1. APPLY 2 COATS SEMI-TRANSPARENT STAIN TO ALL WOOD
   COLOR TO BE SELECTED BY L.A.
2. FOOTING TO BE POURED AGAINST UNDISTURBED SOIL OR RECOMPACTED SOIL PER STRUCTURAL SOILS REPORT.
3. ALL WOOD SHALL BE ROUGH SAWN.
4. COUNTERSINK ALL HARDWARE.

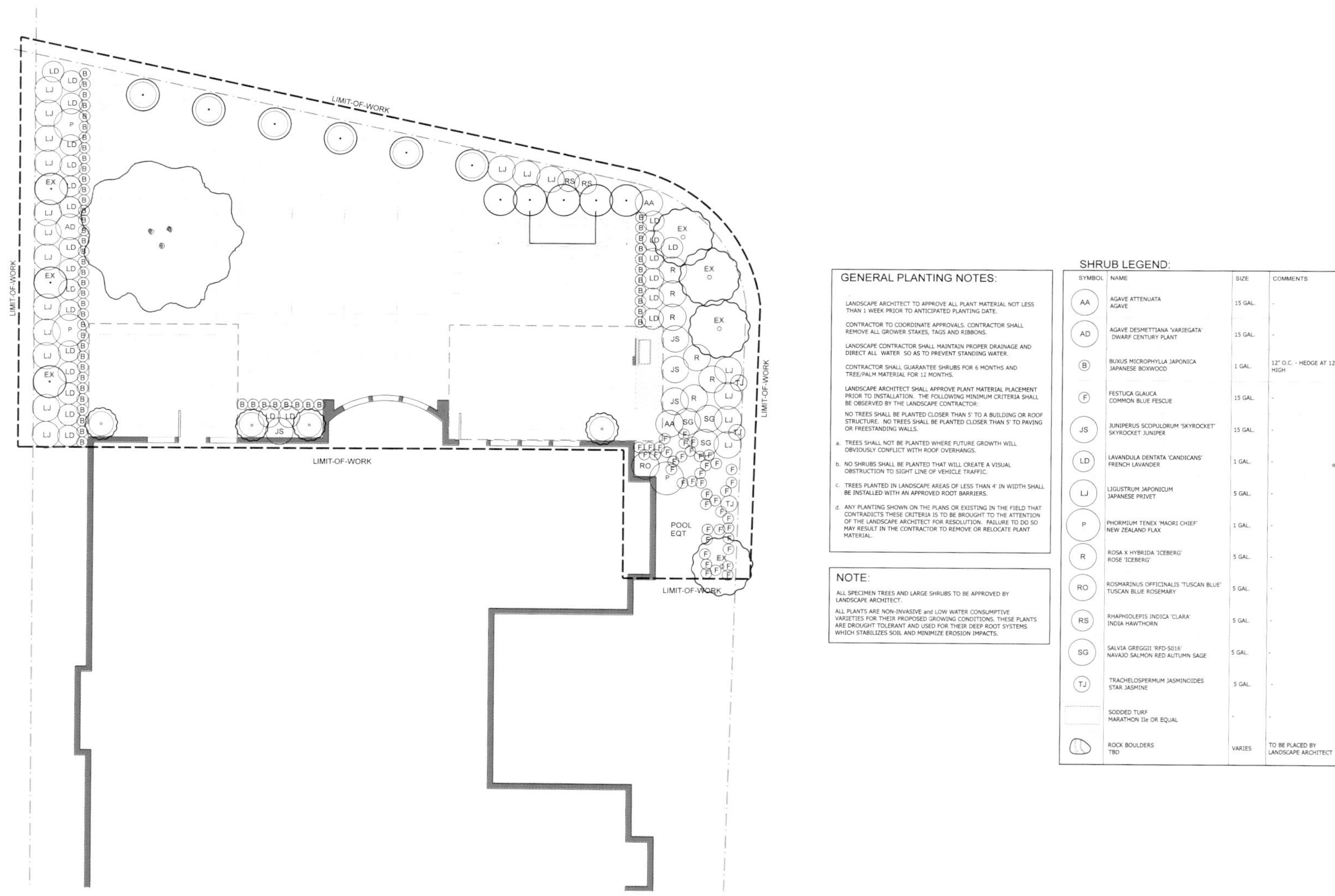

SYMBOL	NAME	SIZE	COMMENTS
AA	AGAVE ATTENUATA AGAVE	15 GAL.	-
AD	AGAVE DESMETTIANA 'VARIEGATA' DWARF CENTURY PLANT	15 GAL.	-
B	BUXUS MICROPHYLLA JAPONICA JAPANESE BOXWOOD	1 GAL.	12" O.C. - HEDGE AT 12" HIGH
F	FESTUCA GLAUCA COMMON BLUE FESCUE	15 GAL.	-
JS	JUNIPERUS SCOPULORUM 'SKYROCKET' SKYROCKET JUNIPER	15 GAL.	-
LD	LAVANDULA DENTATA 'CANDICANS' FRENCH LAVANDER	1 GAL.	-
LJ	LIGUSTRUM JAPONICUM JAPANESE PRIVET	5 GAL.	-
P	PHORMIUM TENEX 'MAORI CHIEF' NEW ZEALAND FLAX	1 GAL.	-
R	ROSA X HYBRIDA 'ICEBERG' ROSE 'ICEBERG'	5 GAL.	-
RO	ROSMARINUS OFFICINALIS 'TUSCAN BLUE' TUSCAN BLUE ROSEMARY	5 GAL.	-
RS	RHAPHIOLEPIS INDICA 'CLARA' INDIA HAWTHORN	5 GAL.	-
SG	SALVIA GREGGII 'RFD-S016' NAVAJO SALMON RED AUTUMN SAGE	5 GAL.	-
TJ	TRACHELOSPERMUM JASMINOIDES STAR JASMINE	5 GAL.	-
	SODDED TURF MARATHON IIe OR EQUAL	-	-
	ROCK BOULDERS TBD	VARIES	TO BE PLACED BY LANDSCAPE ARCHITECT

No.31 YiJia Garden

Landscape Architect
Hothouse Design

Location
Suzhou, Jiangsu Province, China

Area
350 m²

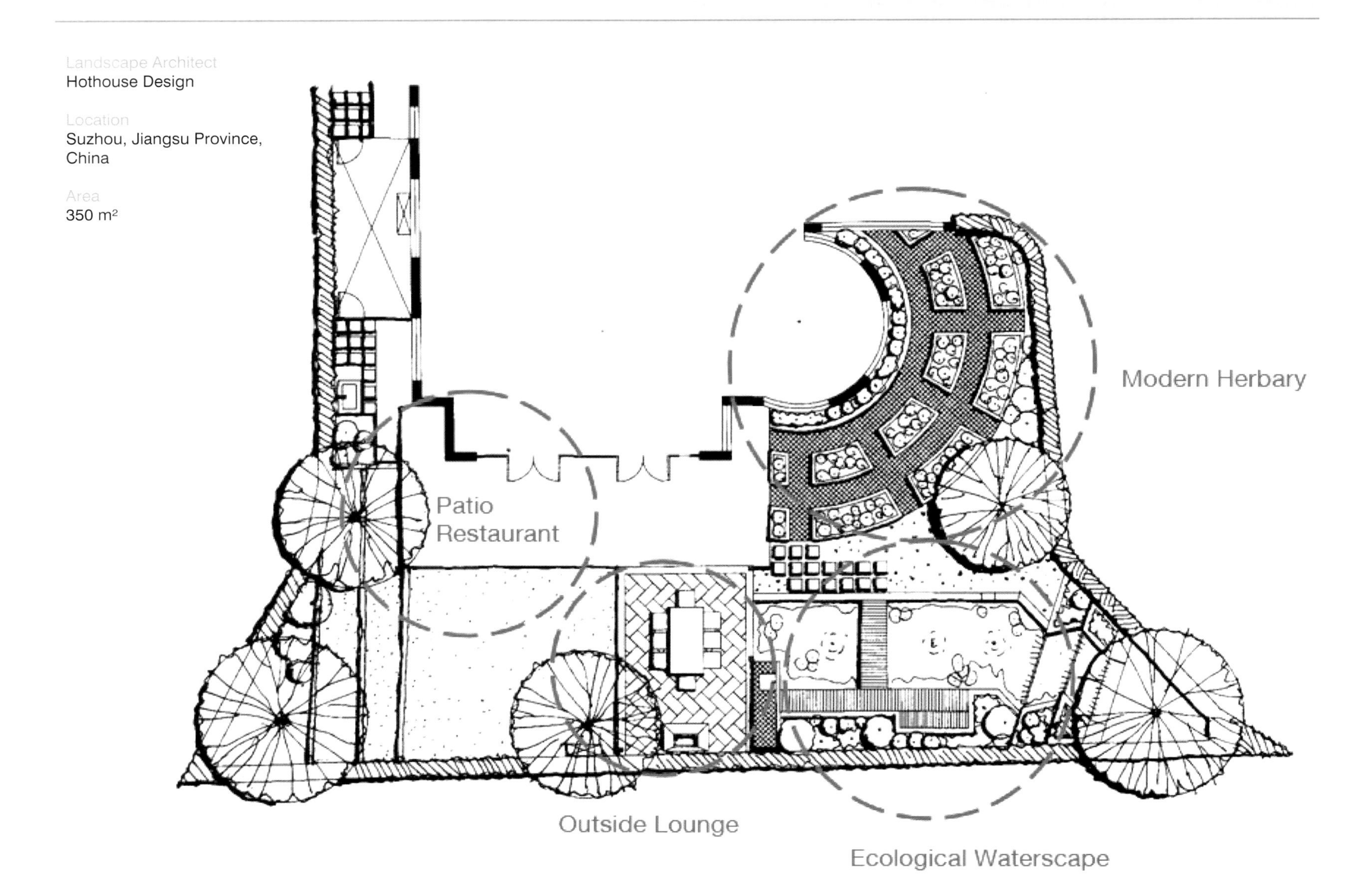

This garden, located on the east of the Zhangjiagang Gymnasium, covers an area of more than 300m². Facing south and enjoying a relatively warm microclimate, it is an ideal location for residence due to its superior geographic position. Mrs.W, the owner of this house, has a great passion for life and a very good taste. She decided to remodel the Villa Garden which was decorated 6 years ago and already outdated. There are not many Villa Gardens in this city, so in the aspect of their designing and constructing, people usually simply plant some trees and spread the lawn. These simple decorations won't meet the owner's expectation and satisfied her trendy aesthetic requirement. Through friend's recommendation, Hothouse Design was commissioned to design her Villa Garden.

The designers still clearly remembered the scene the very first time they entered the graden: It was a long garden, with a length of 20 meters and a width of 6.5 meters. Among the 6.5 meters, 3 meters are covered with glazed ceramic tiles, which makes it more narrow and gives people a sense of discomfort. The only 2 kinds of plants in the garden are tall trees and lawn, and the lawn looked scorched without mowing. Without middle-height plants, the whole space lack a sense of depth and order. An adopted stray dog were running around happily in the garden.

Start with the functionality, the designers applied their design proposition in this project: internalizing the external space. The garden is the extension of indoor living. Initially, they decided to build an outdoor kitchen and dining room, a landscape pool, which corresponds to them. Next, they started to deal with the layout and put the elements they need in this project on papers. The biggest challenge in this project is the unique and narrow site condition of this garden. So the priority is to find out what they can do to resolve the problem.

After repeated discussions and considerations, they decided to solve this problem from the perspective of proportion, which is to divide the whole space into 2 or 3 sections visually.

The designers put the functional area in the east side of the middle area of the garden, put the landscape pool in the east side of the functional area, and lay the lawn in its western side, thus divided the garden according to the proportion of 3:2:5, so the functional area and the landscape pool together occupied half of the whole area, and the other half is covered with the lawn. Thus the final layout and design scheme of the garden formed this way.

Eser Company

Landscape Architect
Dr. Turgay ATEŞ-Vista Urban Design & Landscape Architectural Office

Team
Selma Sarıal, Birgül Özçiçek, Ada Deniz Sel, Nevra Bilge

Client
Eser Construction Company, Ankara

Location
Ankara Çankaya District, Turkey

Photographer
Nepes Atamamedov

The building is a 4-storey building of 777m² located on a land which covers 2,345m² above ground. Building covers 2,345m² area totally underground. Landscape works is done as a roof garden (covering 1,568m²) on the slab of basement floor.

Basic idea behind the design is to try to create an outdoor atmosphere having its own geometry and harmony totally in contrast with the building's rectangular forms, yet following the building macro form orders.

In the design the materials selected for construction and landscaping are all natural, drought tolerant, with less irrigation water demand. These materials are timber, natural stones, pebbles and like. Reflective stainless steel surfaces are functioning to enlarge the volume of the space and as a design trick stands for air, which is again a natural element.

The building is awarded by 'Platinum Leeds certificate' and holds complete 20 points award on landscape design of the related category.

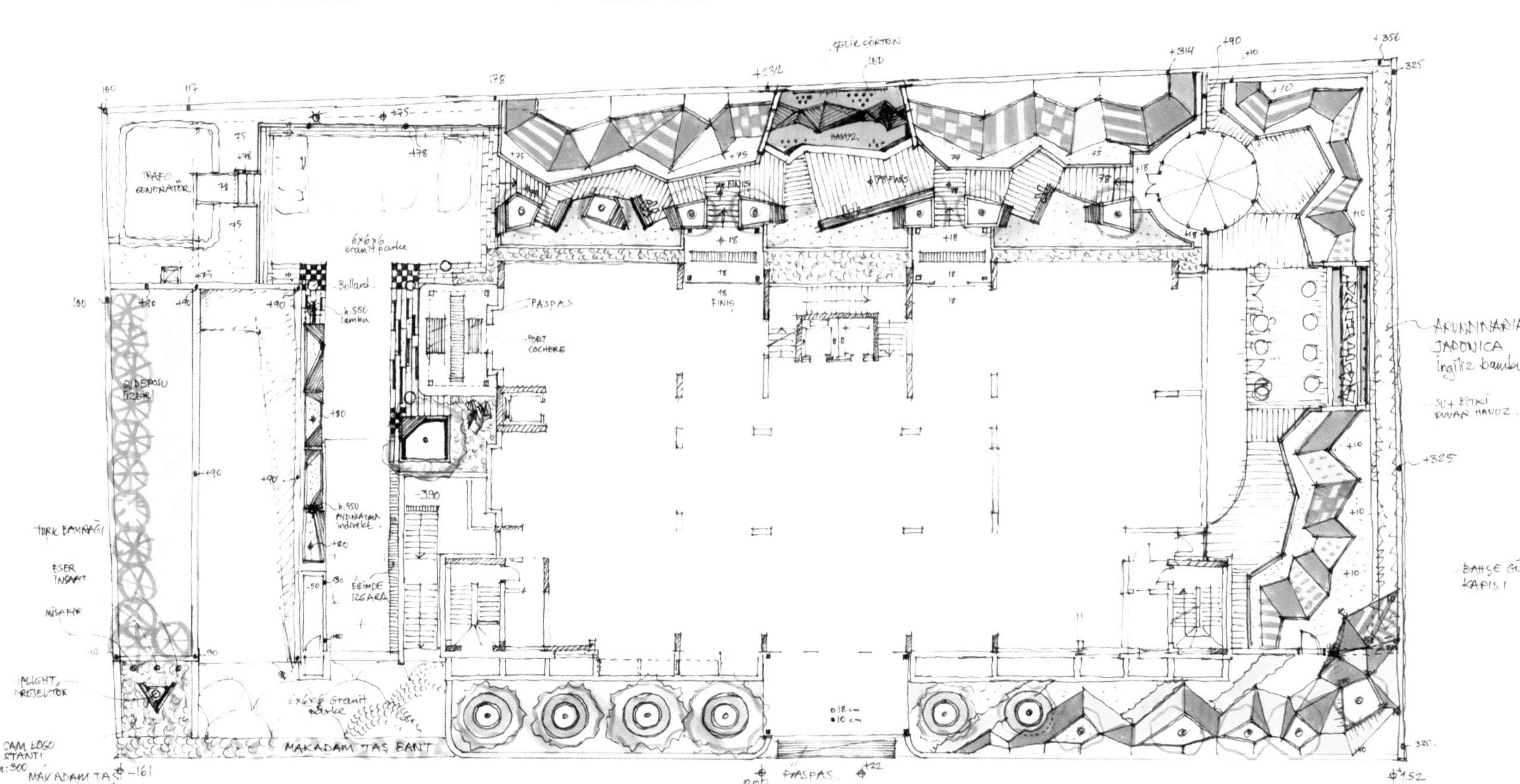

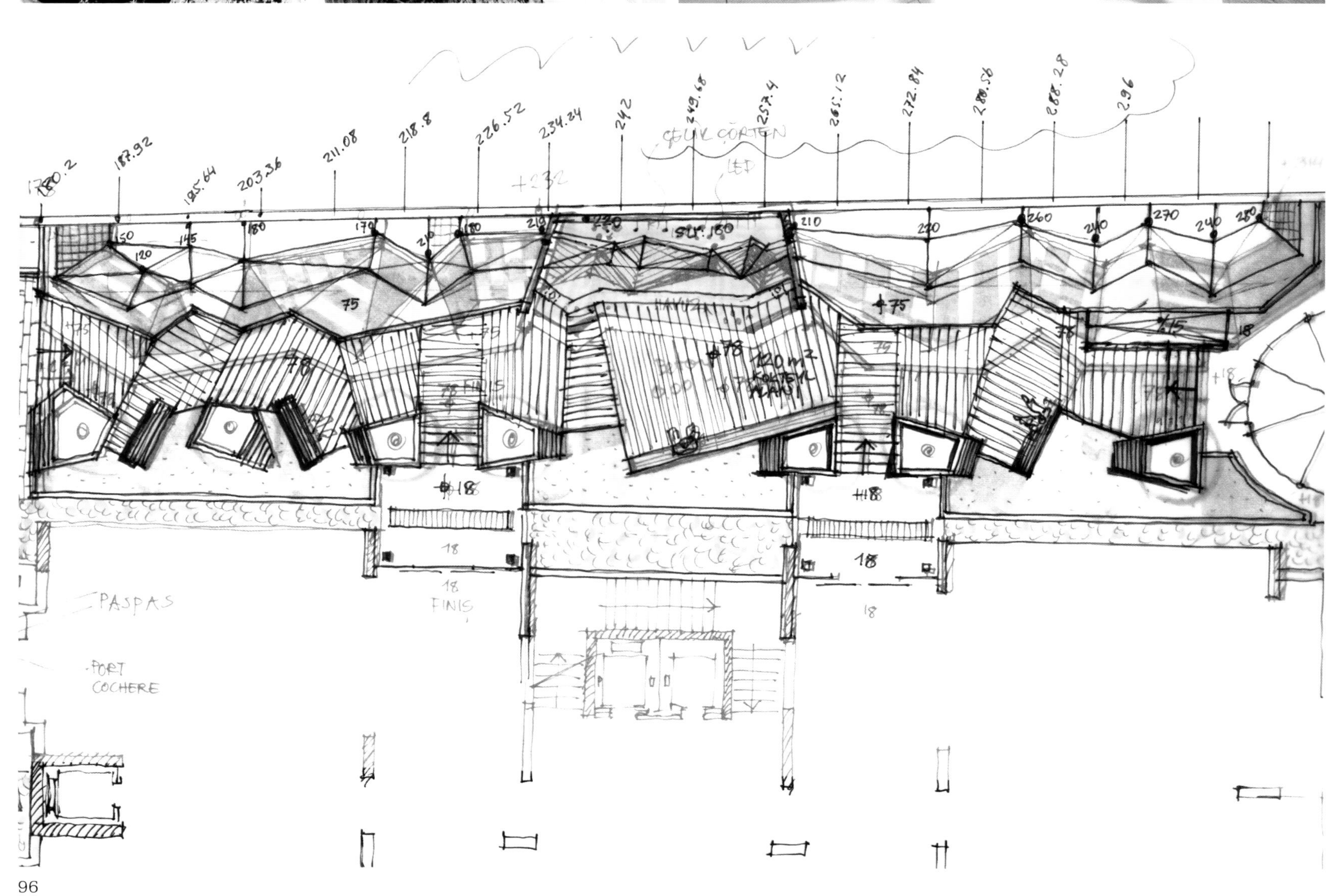
187.92
203.36
211.08
218.8
226.52
234.24
242
249.68
257.4
265.12
272.84
280.56
288.28
296
LED
75
+18
18
PASPAS
FINIS
PORT
COCHERE

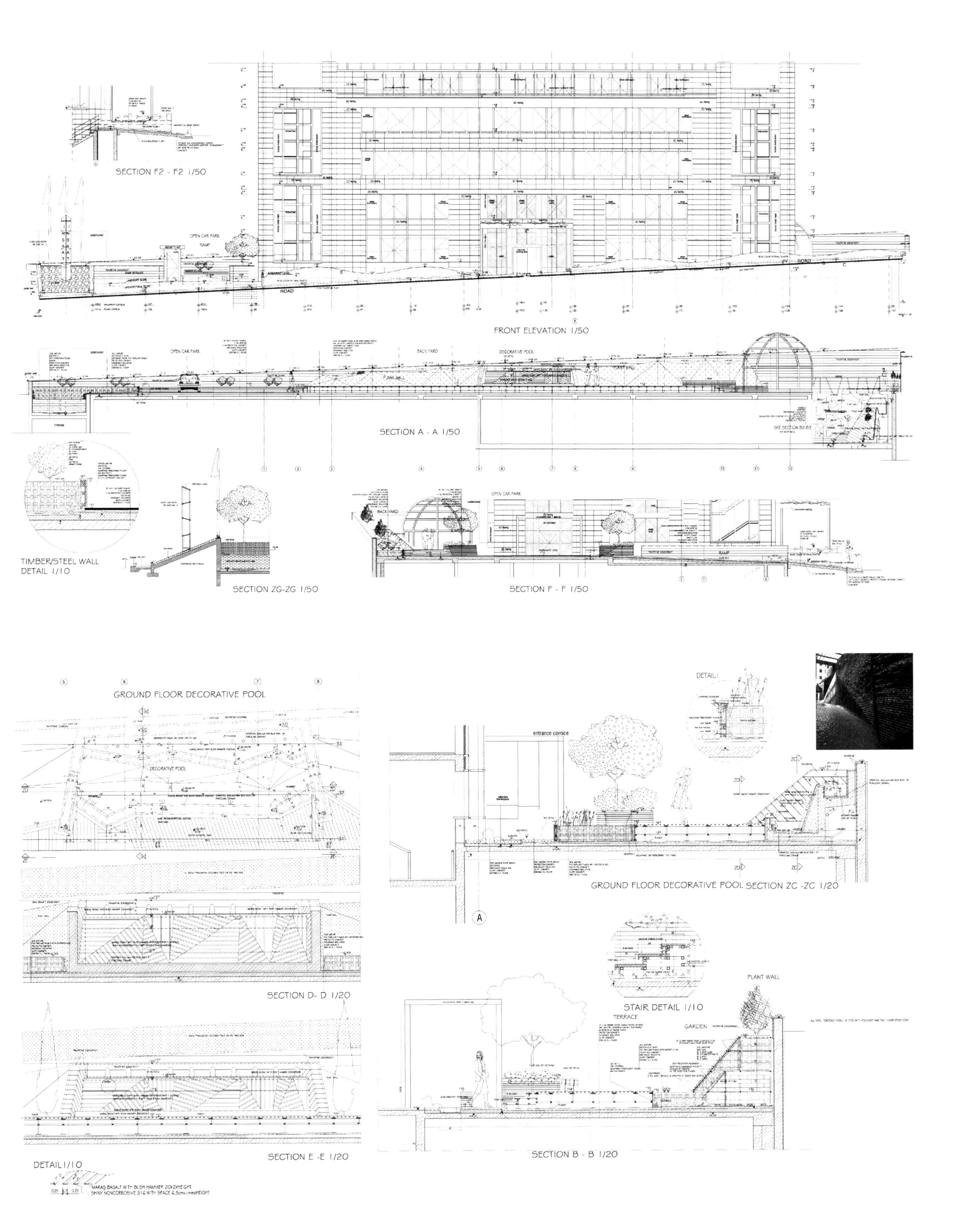
SECTION F2 - F2 1/50
OPEN CAR PARK
RAMP
ROAD
FRONT ELEVATION 1/50
OPEN CAR PARK
BACK YARD
DECORATIVE POOL
PLANT WALL
SECTION A - A 1/50
SEE SECTION B3-B3
TIMBER/STEEL WALL
DETAIL 1/10
SECTION ZG-ZG 1/50
OPEN CAR PARK
BACK YARD
RAMP
SECTION F - F 1/50
GROUND FLOOR DECORATIVE POOL
DECORATIVE POOL
DETAIL I
entrance cornice
GROUND FLOOR DECORATIVE POOL SECTION ZC -ZC 1/20
SECTION D- D 1/20
STAIR DETAIL 1/10
PLANT WALL
TERRACE
GARDEN
SECTION E -E 1/20
SECTION B - B 1/20
DETAIL 1/10
MARAS BASALT WITH BUSH HAMMER 20X2XHEIGHT
SHINY NONCORROSIVE 316 WITH SPACE 6.5cmx1mmxHEIGHT

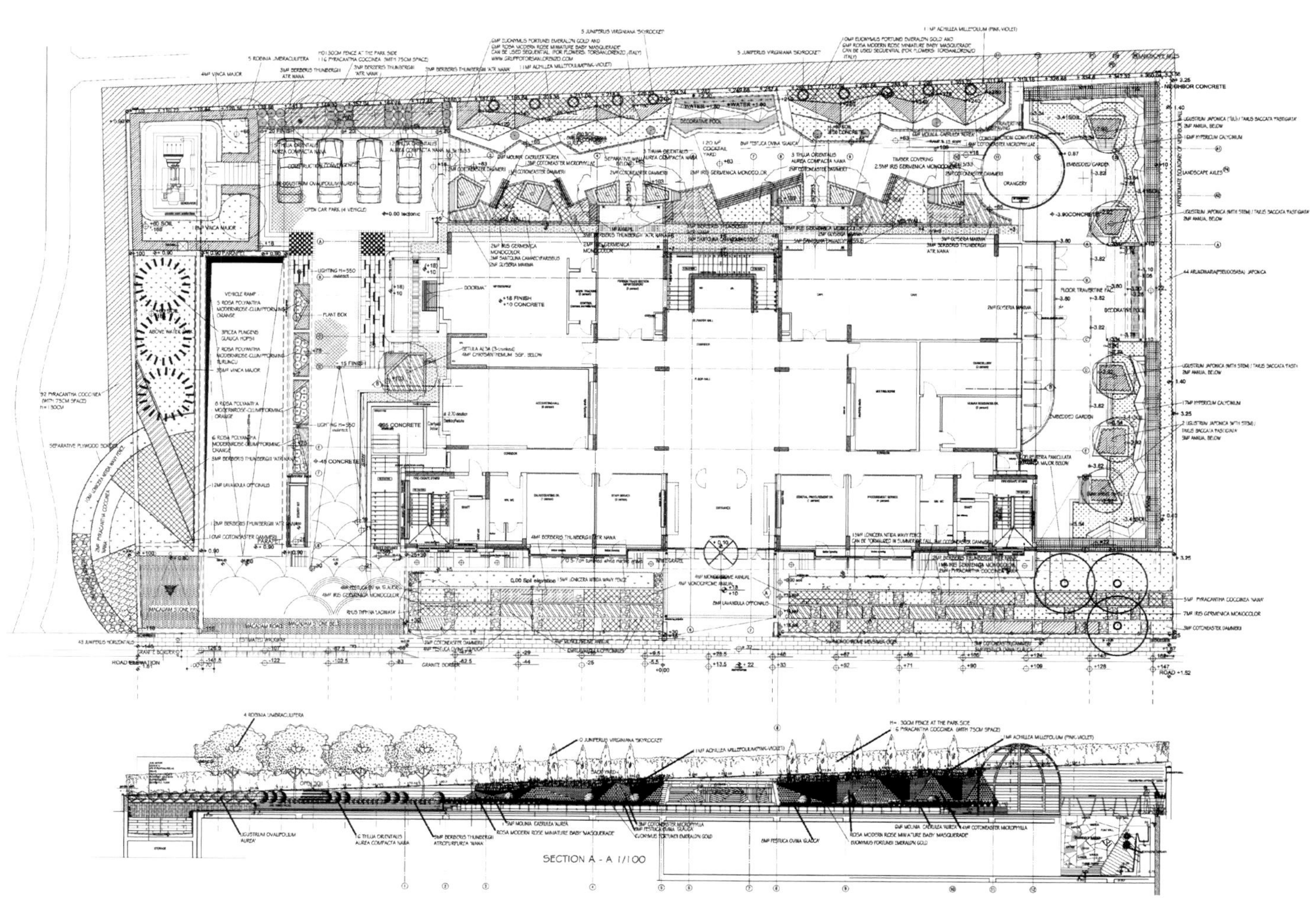
SECTION A - A 1/100

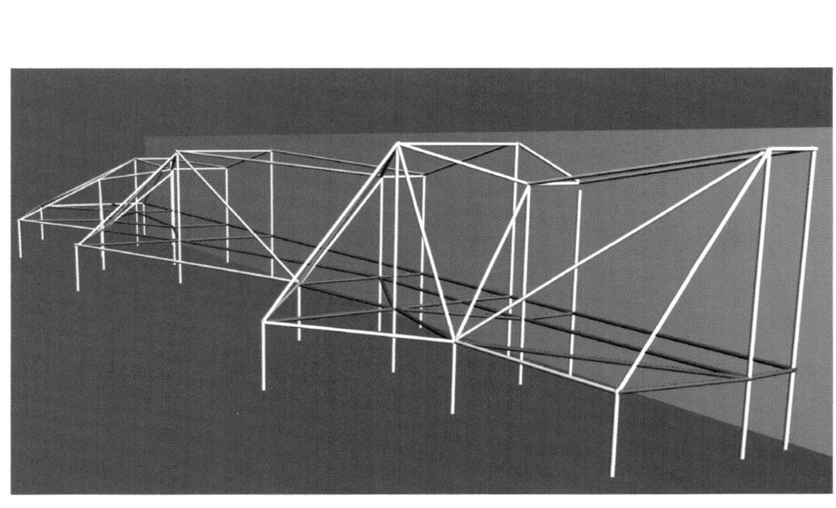

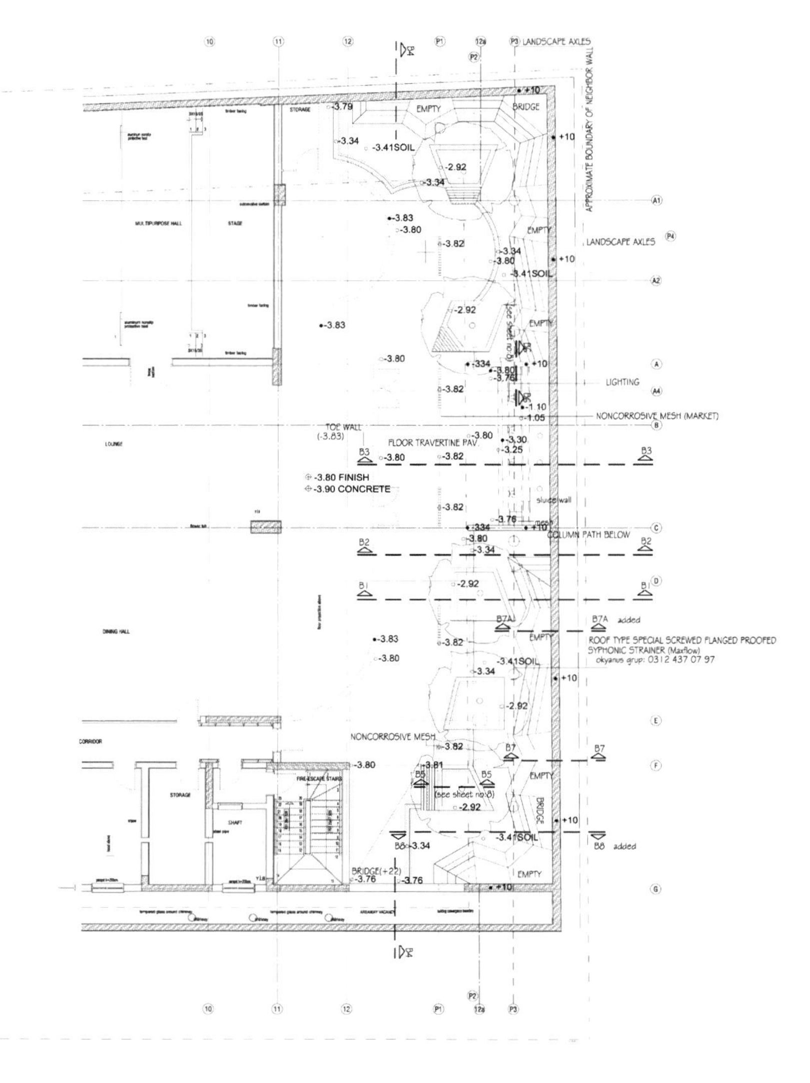

LANDSCAPE AXLES
APPROXIMATE BOUNDARY OF NEIGHBOR WALL
EMPTY
BRIDGE
-3.79
-3.34
-3.41SOIL
-2.92
-3.34
-3.83
-3.80
-3.82
LANDSCAPE AXLES
-3.41SOIL
+10
-2.92
LIGHTING
NONCORROSIVE MESH (MARKET)
TOE WALL
(-3.83)
FLOOR TRAVERTINE PAV
-3.80 FINISH
-3.90 CONCRETE
COLUMN PATH BELOW
ROOF TYPE SPECIAL SCREWED FLANGED PROOFED
SYPHONIC STRAINER (Maxflow)
okyanus grup: 0312 437 07 97
added
NONCORROSIVE MESH
FIRE ESCAPE STAIRS
STORAGE
(see sheet no.8)
BRIDGE(+22)
-3.76

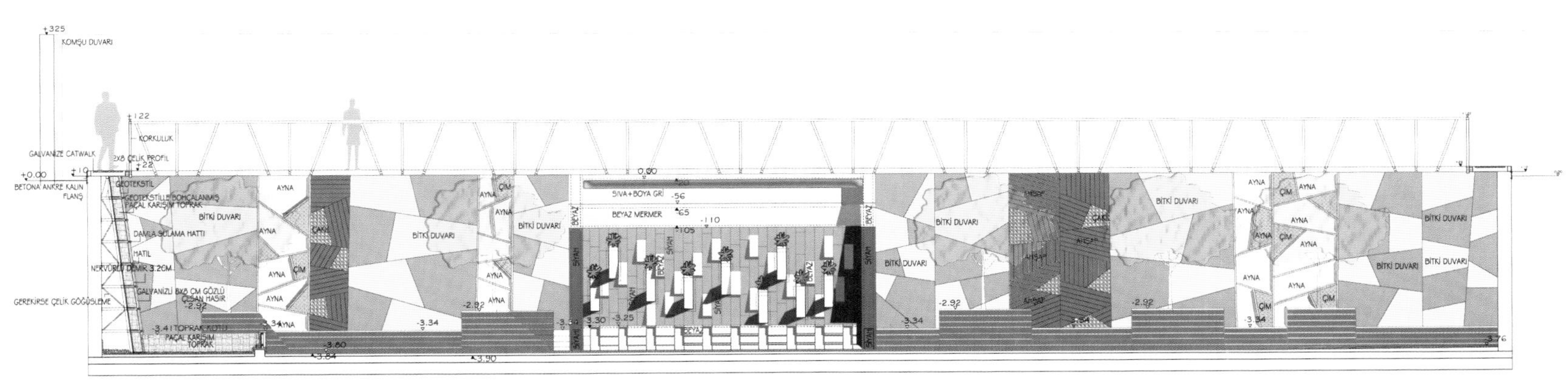
KOMŞU DUVARI
GALVANİZE CATWALK
KORKULUK
BETONA ANKRE KALIN FLANŞ
GEOTEKSTİL
AYNA
BİTKİ DUVARI
ÇİM
SIVA+BOYA GRİ
BEYAZ MERMER
DAMLA SULAMA HATTI
GERENİRSE ÇELİK GÖĞÜSLEME

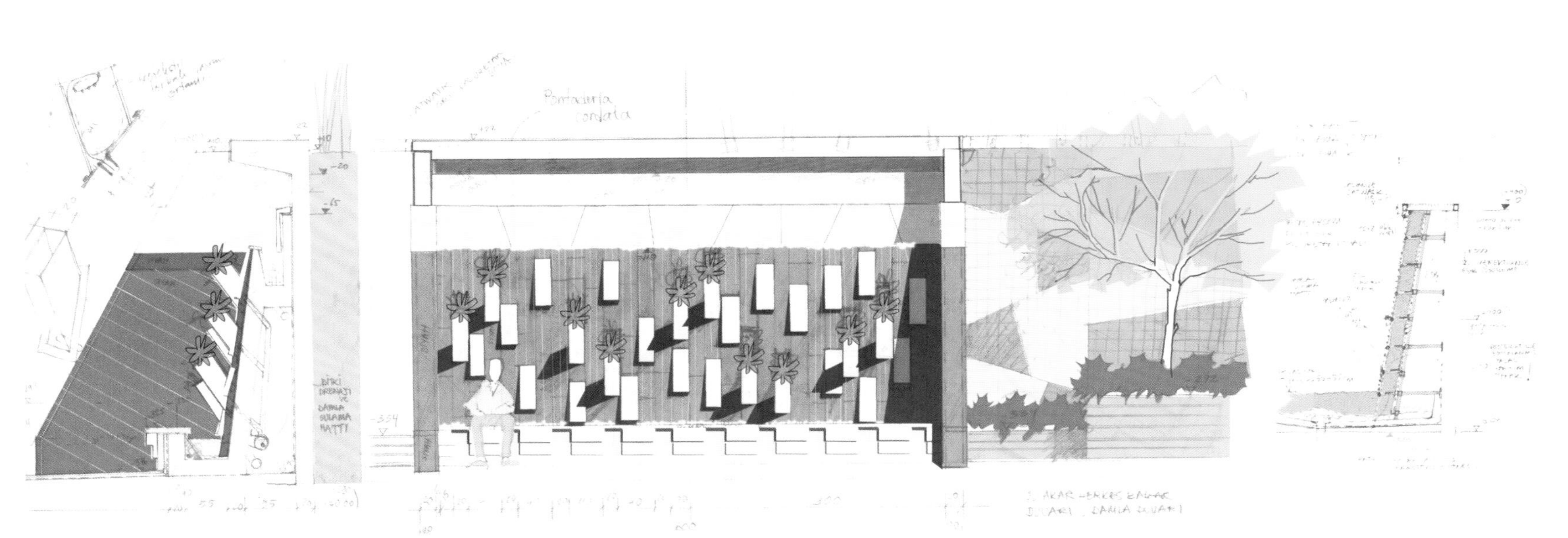

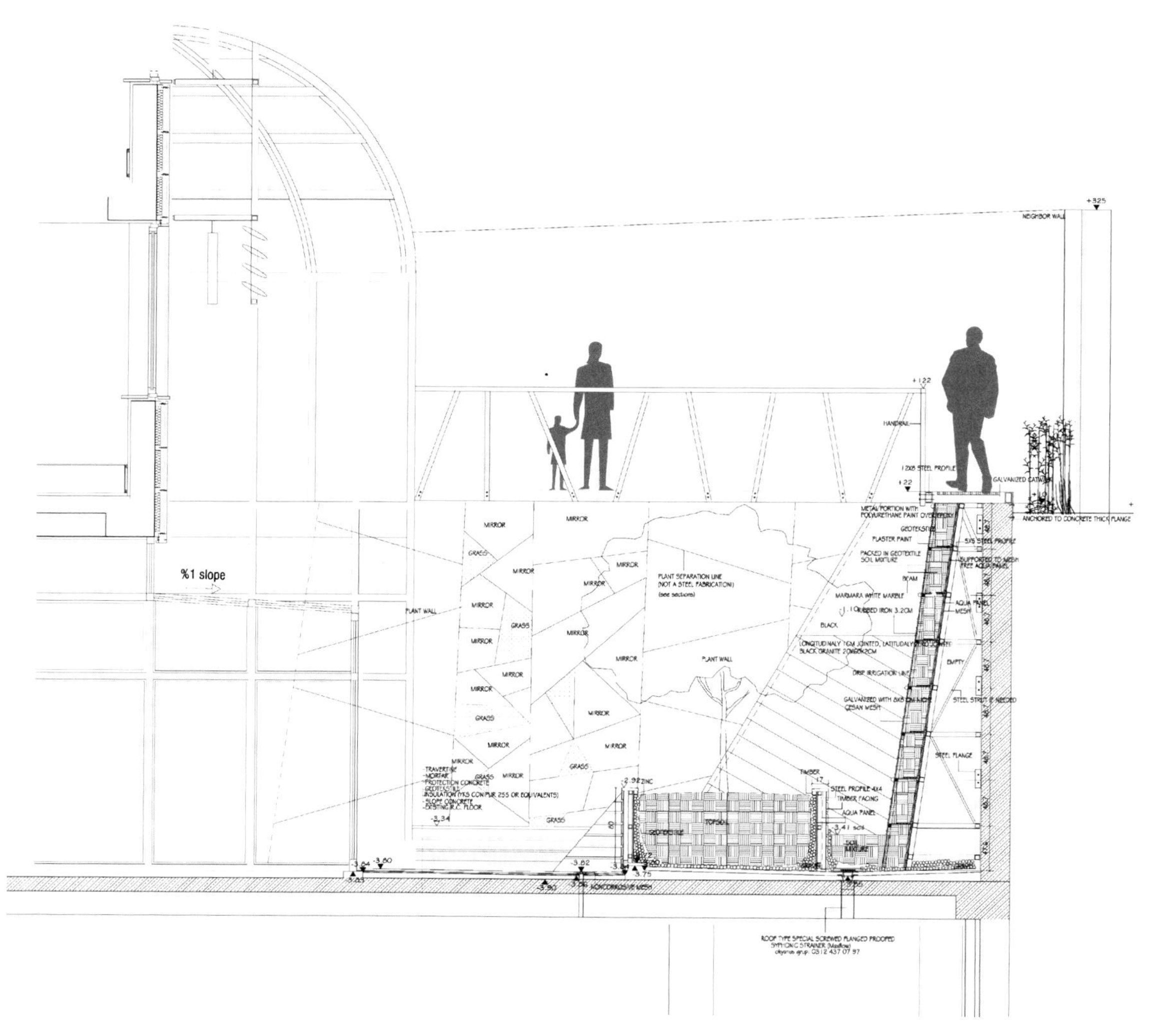

SECTION B1 - B1 1/20

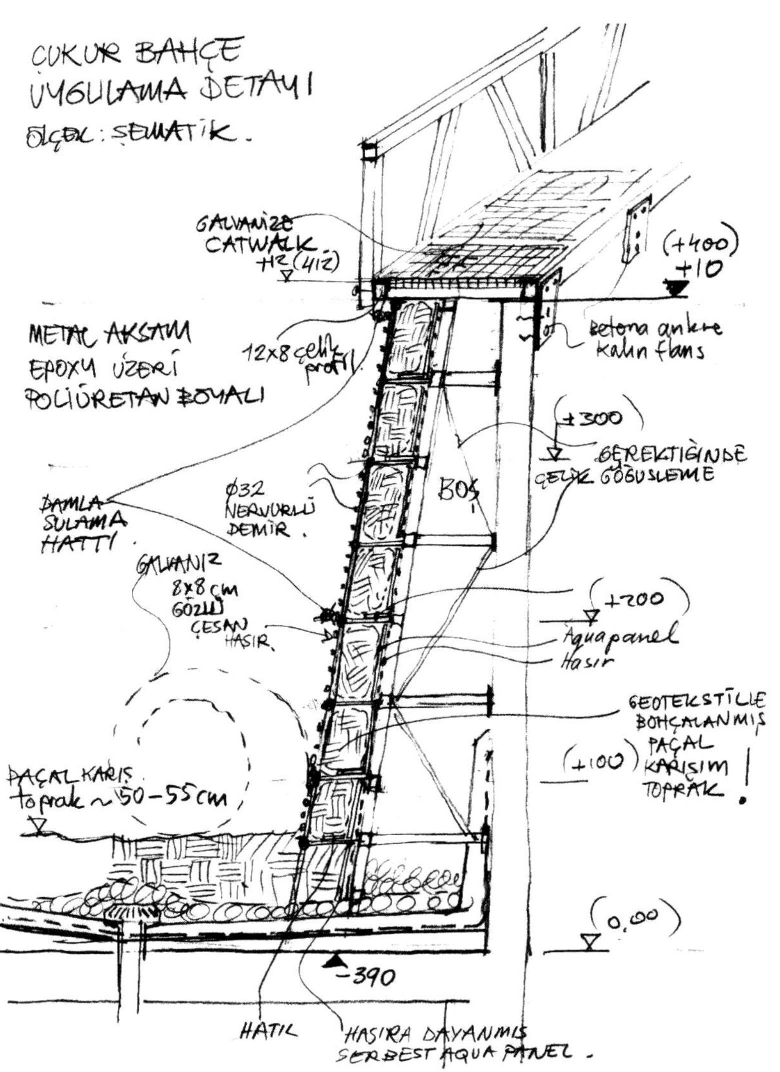

ÇUKUR BAHÇE
UYGULAMA DETAYI
ÖLÇEK: ŞEMATİK.
GALVANİZ CATWALK
+12(412)
(+400)
+10
Betona ankre kalın flanş
METAL AKSAM
EPOXY ÜZERİ
POLİÜRETAN BOYALI
12x8 çelik profil
(+300)
GEREKTİĞİNDE ÇELİK GÖĞÜSLEME
BOŞ
DAMLA SULAMA HATTI
Ø32 NERVÜRLÜ DEMİR
GALVANİZ 8x8 cm GÖZLÜ ÇESAN HASIR
(+200)
Aquapanel
Hasır
GEOTEKSTİLLE BOHÇALANMIŞ PAÇAL KARIŞIM TOPRAK !
(+100)
PAÇAL KARIŞ. toprak ~50-55cm
(0,00)
-390
HATIL
HASIRA DAYANMIŞ SERBEST AQUA PANEL.

Trenton Drive

Landscape Architect
Ecocentrix Landscape Architecture

Area
2,023 m^2

Location
California, USA

Photographer
John Feldman

The designers were charged with reinventing the gardens for this Colonial Revival style home in Beverly Hills, California. What was perhaps the greatest challenge was to orient a clean-lined, classical estate garden around an existing oddly shaped swimming pool. By incorporating herringbone and basket weaved brick patterns, along with subtle inclusion of 'X' patterns in other paved details and site accessories, they were able to fashion this pool into something of 'focal piece'.

While classical by order, there is a distinct use of plant materials that within the boxwood edged planter beds is still very 'California' by nature. There is a purposeful overlay of contemporary space making in the creation of this classical garden. Clean lines and distinct forms strongly cue to the orthogonal architecture. The 'dogleg' shaped pool was a hurdle that was overcome by strong design allegiance to symmetry and axial circulation through the garden.

Casual crushed gravel pathways relent a wonderful sound under foot, and carry through to the dining garden. Brick paving details were used in other areas to create intended dialogue with lawn planted joints found in concrete paving and also to mimic the branded 'X' on the black enamel painted Versailles planters.

Iron privacy rails adorn the tops of white washed brick walls, while a trellis, fences, and gates, are boldly crafted from milled lumber. They all are fashioned to reflect and enhance the inherent details found in this Classical Beverly Hills home's architecture. The context of built and living details is congruent with its architecture and ground this home that is steeped in rich history and tradition.

Ng Residence

Landscape Architect
ASPECT Studios

Architect
Marsh Cashman Koolloos Architects

Location
Sydney, New South Wales, Australia

Area
362 m²

Photographer
Simon Wood

Awards
2011 Wilkinson Award (NSW Architecture Awards)

A new residence in Sydney is a unique fusion between architecture and landscape.

The intent of the design is to blur the line between internal and external space. Floor surfaces extend from inside the home out into the garden spaces.

The external spaces are separated into three main areas: the street frontage, the central courtyards, and the rear garden space. The front garden comprises small deciduous trees with understorey planting and a gravel courtyard at the house entry. The planting philosophy focused on selection of species that are suited to the microclimate.

All species are drought tolerant and hardy once established. Water is to be recycled from the roof and stored in a concealed tank at the front of the property. Landscape materials are to be 'true materials', a robust and simple palette that are durable and minimise the need for complex maintenance regimes.

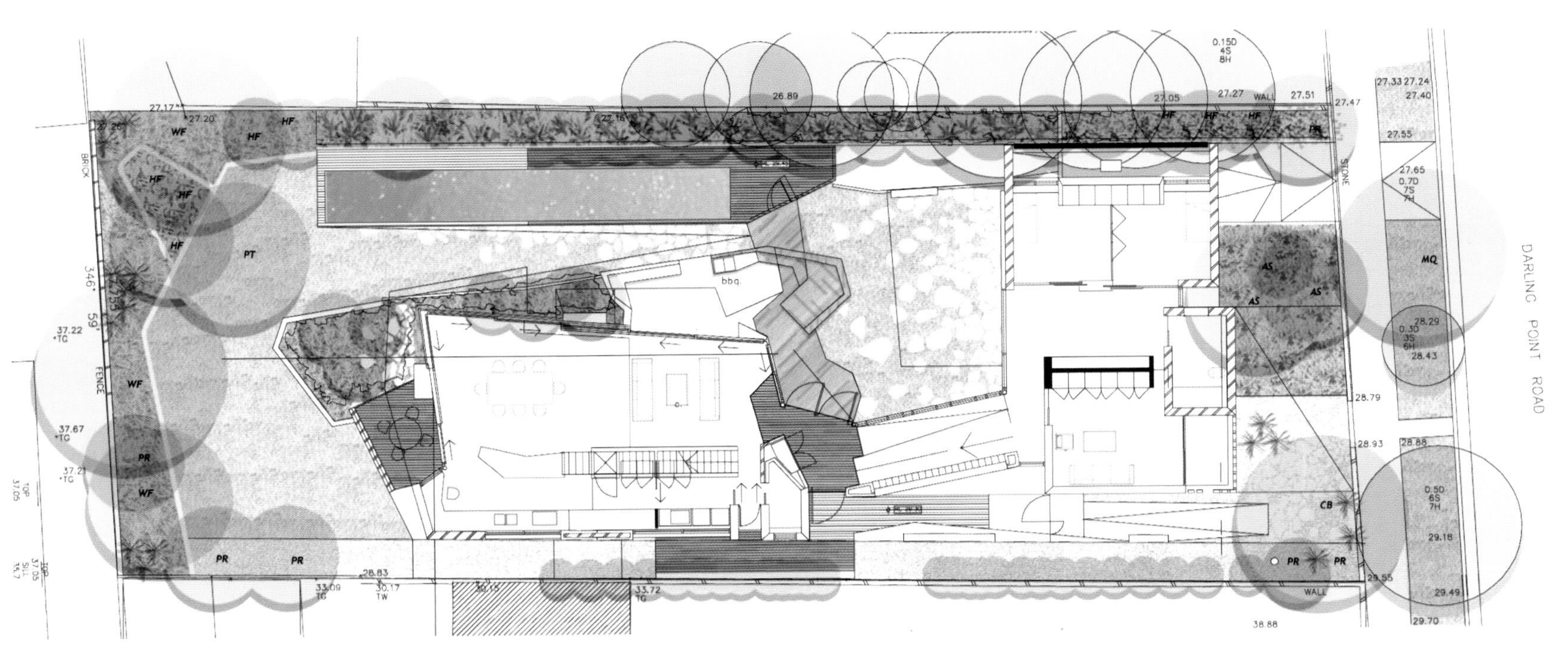

DARLING POINT ROAD
bbq.
WALL
FENCE
BRICK
STONE
WF
HF
PT
PR
AS
MQ
CB

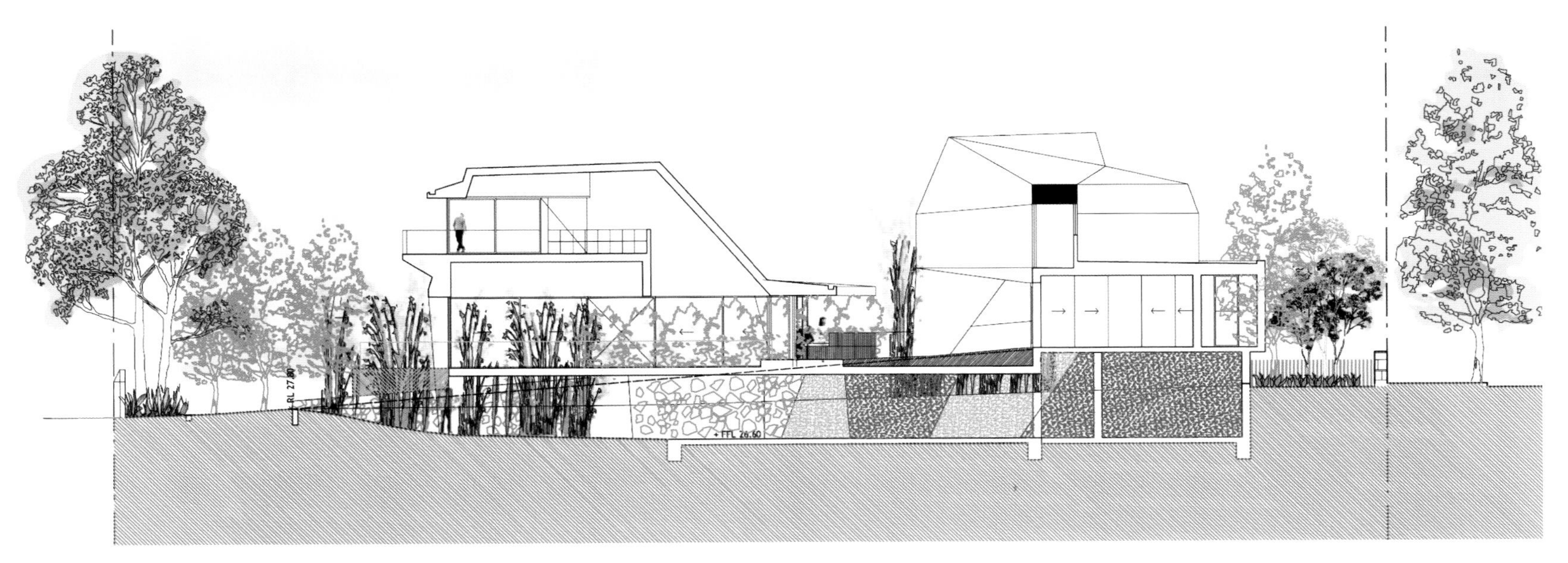

Baja Modern

Landscape Architect
Lewis Aqüi Landscape + Architectural Design, LLC

Location
Plantation, Florida, USA

Photographer
Lewis Aqüi Photography

This new residential masterpiece was a design collaboration at its ultimate. The concept of the Mexican Riviera was implemented in the design and material selections of the hardscape elements and garden features of the estate. The program included a welcoming motor courtyard, playful water features, a dynamic swimming pool, an outdoor shower concealed behind a fluidly moving garden wall, and an outdoor kitchen beneath the shade of a Mexican 'palapa'.

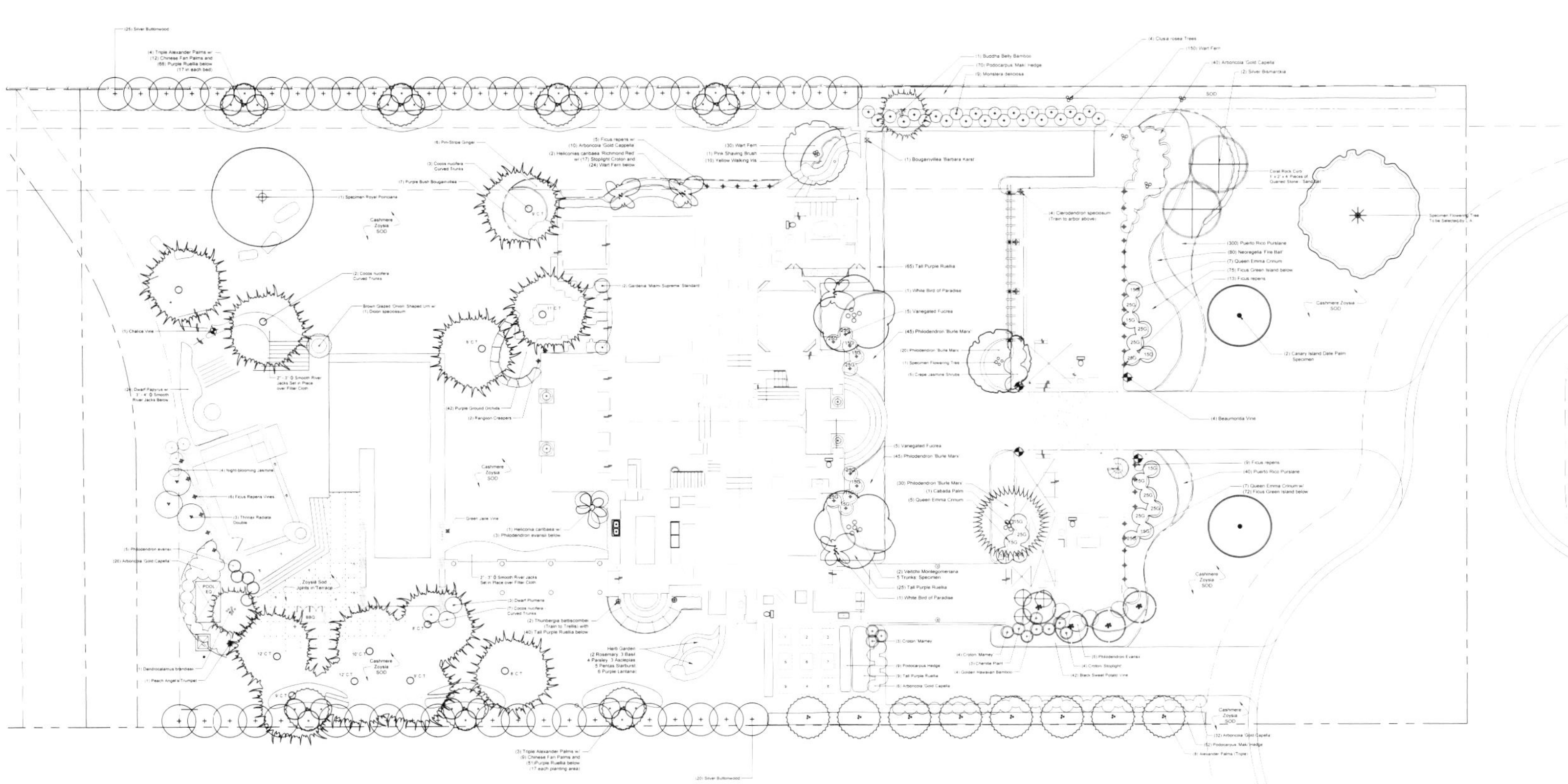

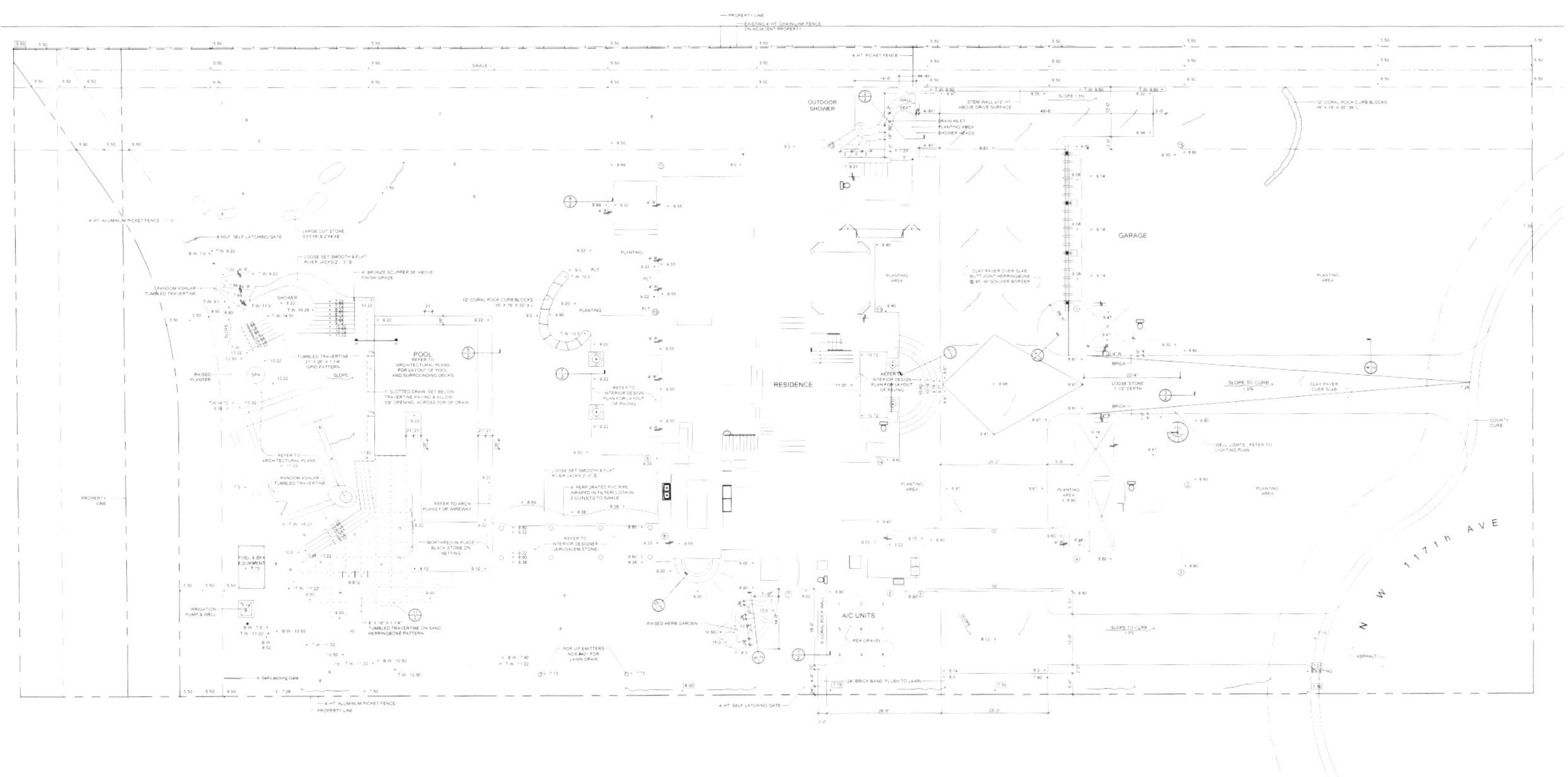
OUTDOOR SHOWER
GARAGE
POOL
RESIDENCE
A/C UNITS
N W 117th AVE

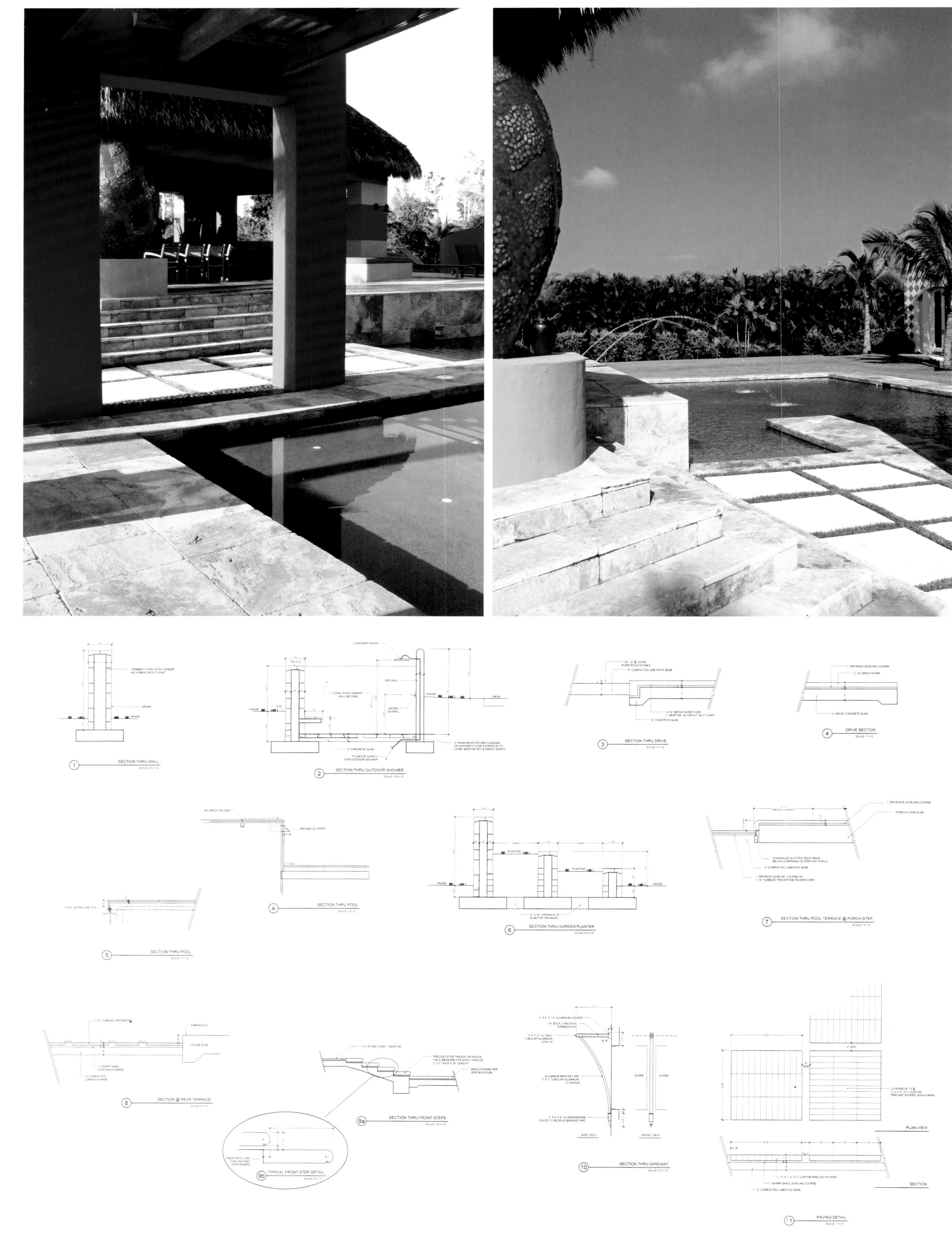
SECTION THRU WALL
SECTION THRU OUTDOOR SHOWER
SECTION THRU DRIVE
DRIVE SECTION
SECTION THRU POOL
SECTION THRU POOL
SECTION THRU GARDEN PLANTER
SECTION THRU POOL TERRACE @ PORCH STEP
SECTION @ REAR TERRACE
SECTION THRU FRONT STEPS
TYPICAL FRONT STEP DETAIL
SECTION THRU WIREWAY
PLAN VIEW
SECTION
PAVING DETAIL

Smoorenberg

Landscape Architect
Scott Brown Landscape Design

Pool
Custom-built, reinforced concrete shell, fully tiled interior

Retaining Walls
Masonry – fired clay bricks

Pool Coping and Paving
Concrete pavers

Wall Capping
Concrete pavers

Pergola
Timber and treated pine beams

Lighting
Low-voltage halogen
Megabay and Lumascape

Water Feature
Fired clay urn

Main Feature Tree
Canary Island Date Palm

Hedge Planting around Pool
Lilly-pilly

Although this project began with some pre-existing elements within the backyard, the prospect of integrating those which the designer wanted to preserve made the design project a significant challenge. Starting with a blank canvas is usually much easier, but this was not feasible in this case.

The designer decided the Canary Island Date Palm could be such a great asset, however its existing position meant that it was almost a metre 'off centre' as a focal point from the main central hallway which bisects the house from the front doors to the rear, bay window of the meals area.

Likewise the pool, tucked in one corner and completely isolated, had no relationship with the home.

The 15.24m Camphor laurel tree, originally dominating the yard needed to go – in order to provide the client with the structured outdoor space they were looking for. Likewise the isolated, angled deck beside the pool.

The resulting design aims to create an outdoor 'precinct' to the home. A precinct where all the required elements, such as the swimming pool, the outdoor dining area, the sun baking area, the lawn covered recreation area for the kids, the off centre 'focal point' (the Palm) all related to each other and the house in a harmonious way. This themed relationship needs to be completed on two levels. The first is physical, because the areas need to infinitely useable. The other level is visual. This second and more abstract relationship enables the outdoor environment to be enjoyed day or night, rain, hail or shine from inside the home looking out. The various 'sub-rooms' of the outdoor environment fit together like pieces of a jig saw puzzle that extend the interior of the home outwards. This relationship continues when one is outside. By dividing the yard into these 'sub - rooms', more interest and drama is created and intimacy is also created as one feels as if he is within a smaller, intimate outdoor room, which is part of a series of other outdoor rooms – all of which can be seen or glimpsed whilst sitting beside the pool of dining in the barbeque area.

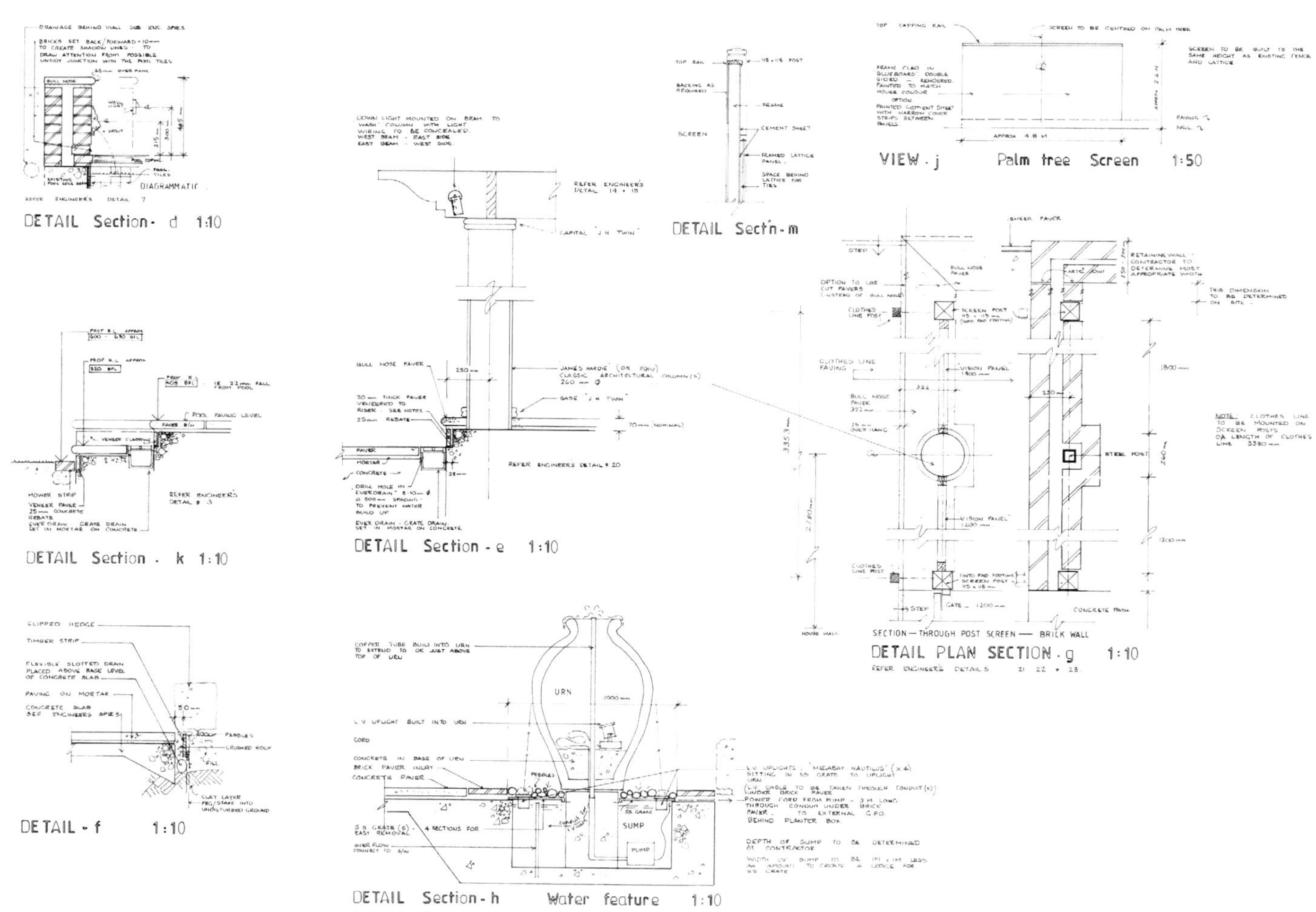

DETAIL Section- d 1:10
VIEW . j Palm tree Screen 1:50
DETAIL Sect'n - m
DETAIL Section . k 1:10
DETAIL Section - e 1:10
SECTION — THROUGH POST SCREEN — BRICK WALL
DETAIL PLAN SECTION . g 1:10
URN
SUMP
PUMP
DETAIL - f 1:10
DETAIL Section - h Water feature 1:10

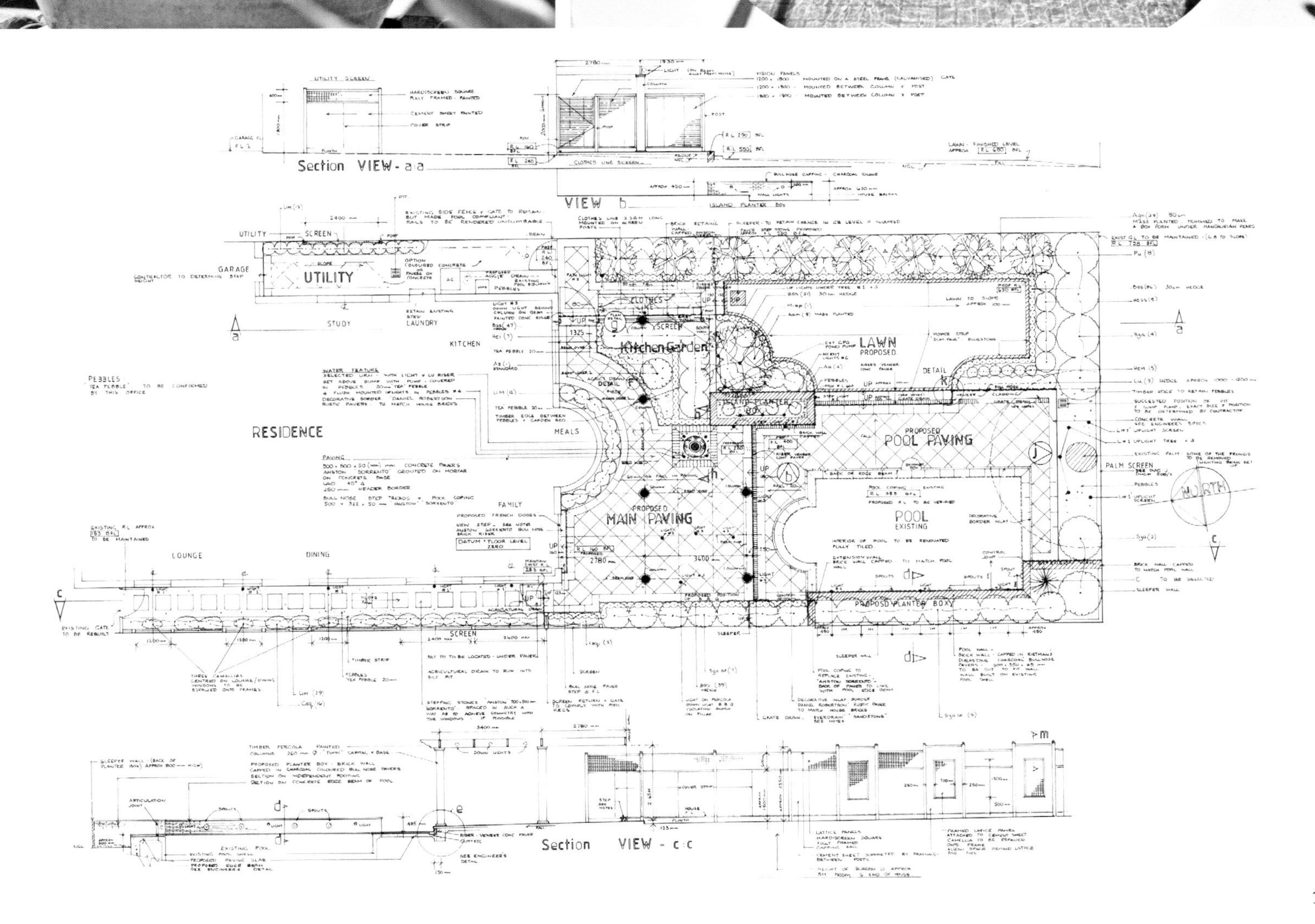

Section VIEW-a a
VIEW b
UTILITY
SCREEN
GARAGE
STUDY
LAUNDRY
KITCHEN
Kitchen Garden
LAWN
PROPOSED
DETAIL
RESIDENCE
MEALS
PROPOSED
POOL PAVING
PALM SCREEN
NORTH
FAMILY
PROPOSED
MAIN PAVING
POOL
EXISTING
LOUNGE
DINING
PROPOSED PLANTER BOX
SCREEN
Section VIEW - c c

Black & White Contemporary House Garden

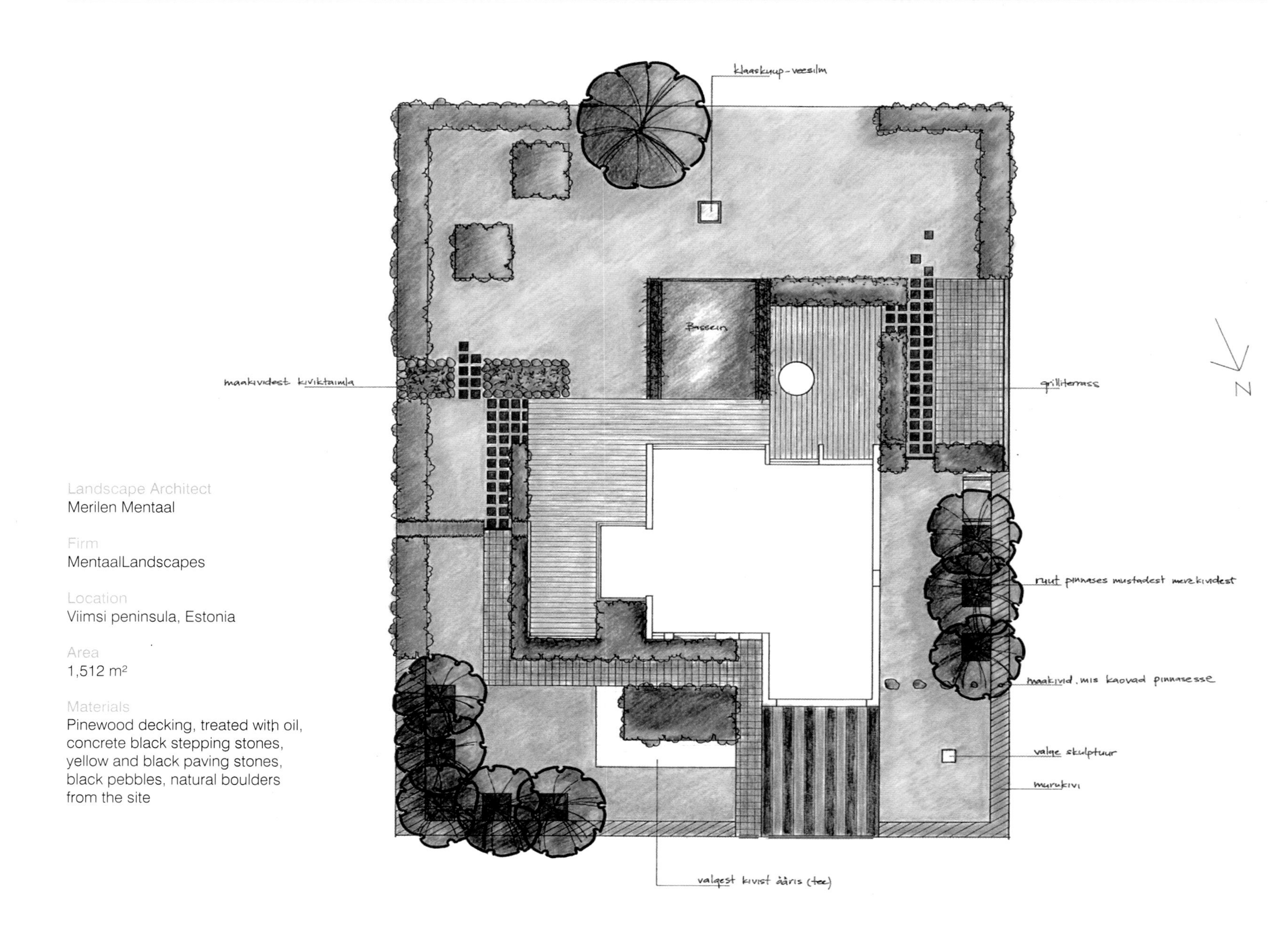

Landscape Architect
Merilen Mentaal

Firm
MentaalLandscapes

Location
Viimsi peninsula, Estonia

Area
1,512 m²

Materials
Pinewood decking, treated with oil, concrete black stepping stones, yellow and black paving stones, black pebbles, natural boulders from the site

Partly hidden within a coniferous forest, this garden of a modern, large black and white family house was ideal for a geometric design, clear, straight lines and plenty of space to live within and move around, softened by lush, generous planting. Large terrace area which starts from one side garden and reaches another side garden, divides outdoor living space into a morning coffee terrace in the east, to sun-terrace in the back and to dining area in the evening sun.

Adding a bit colour to the black and white colour scheme, yellow paving stone was used for main garden path and black paving stone for driveway tieing it with the house. Blackish stepping stones continue from the front of rockery gate into back garden and terraces.

Playing with geometric shapes, an L-shaped area was introduced in the front garden, filled with low grasses (Festuca gautierii, Festuca glauca) edged by black pebbles. Black, polished granite house wall next to the main entrance that has plentiful soft planting in front, creates exciting reflections and allows the white blooms to shine in the morning light.

Large squares of black pebbles along the garden boundary, with Japanese cherries Prunus serrulata 'Kanzan', richly decorate the spring garden with their soft pink blooms.

A round metal fire pit within the ground close to dining terrace allows a safe but natural looking log-fire to be lit in the garden. West-facing side garden was large enough to add raised wooden plants to grow herbs and vegetables that were not of obvious importance when designing the garden but seemed absolutely necessary when the children (twins) were born, about a year after the completion of the garden. Luckily there was plenty of playful garden space left to be filled with swings, slides and everything that comes with the dear little ones.

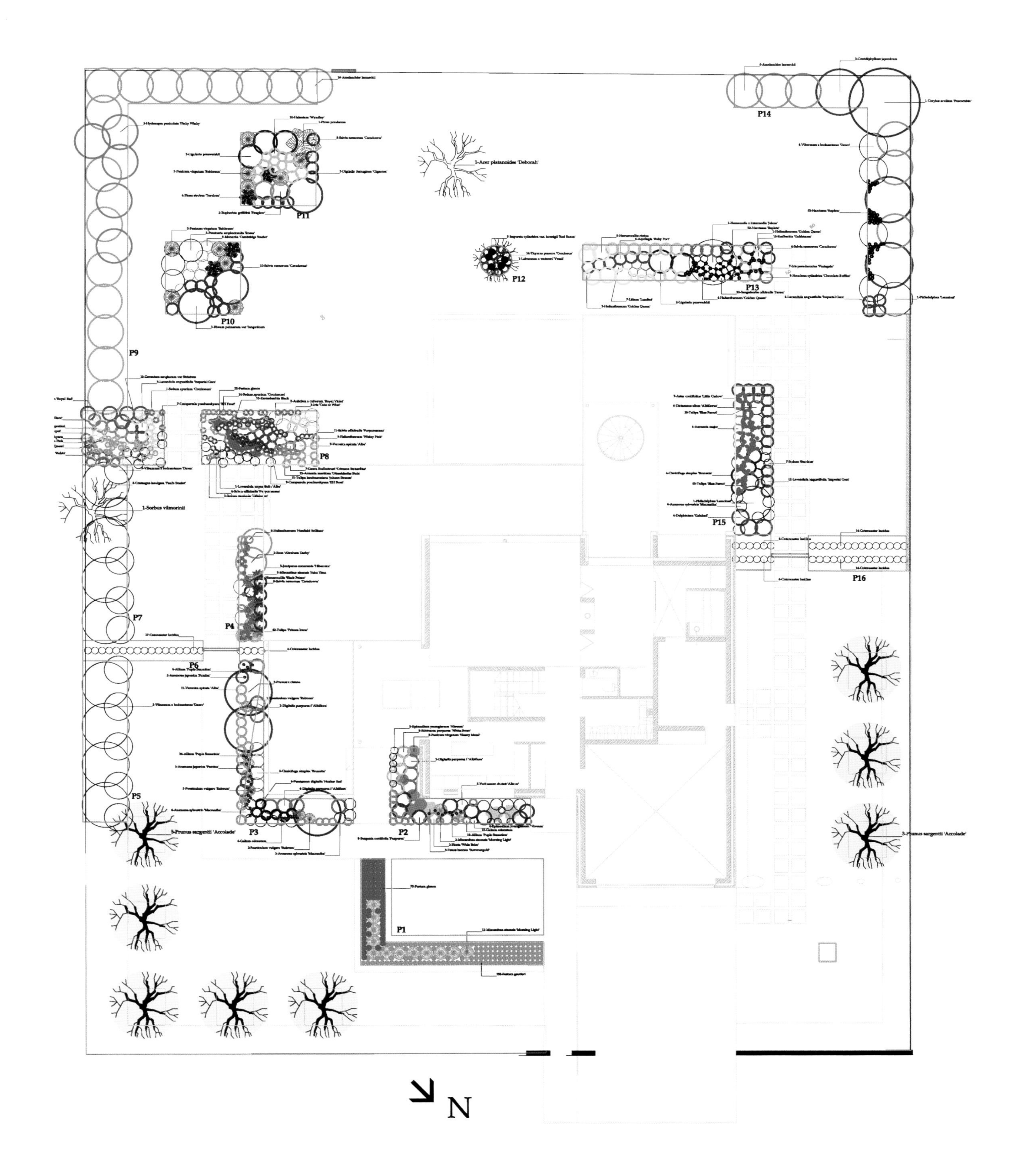

1-Acer platanoides 'Deborah'
1-Sorbus vilmorinii
5-Prunus sargentii 'Accolade'
3-Prunus sargentii 'Accolade'
P1
P2
P3
P4
P5
P6
P7
P8
P9
P10
P11
P12
P13
P14
P15
P16
N

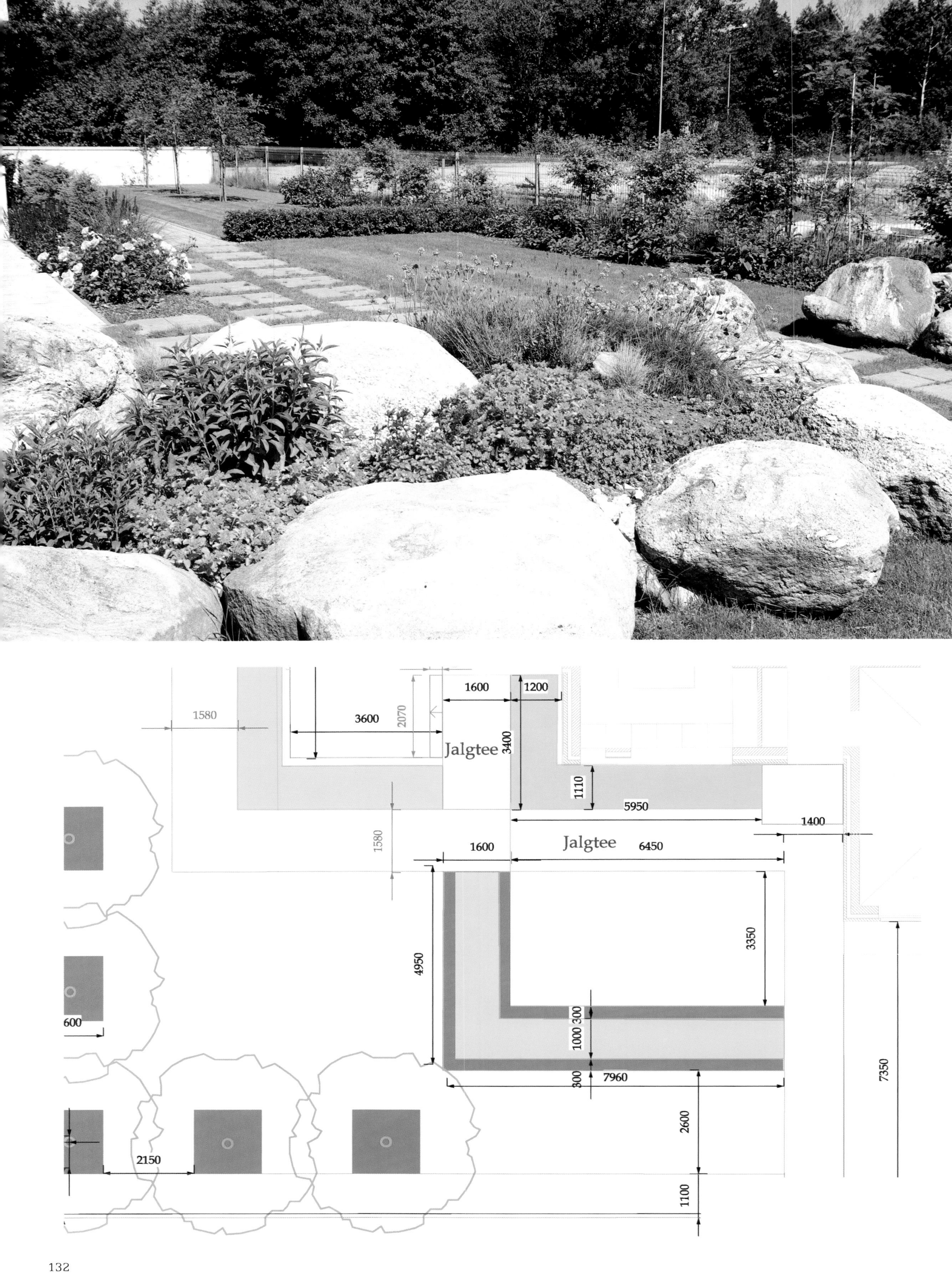
1600
1200
1580
3600
2070
Jalgtee
3400
1110
5950
1400
1580
1600
Jalgtee
6450
4950
3350
300
1000
300
7960
600
7350
2600
2150
1100

From Festival Garden into Tartu Garden

Landscape Architect
Merilen Mentaal – MentaalLandscapes & Maarja Gustavson – Polka OÜ

Location
Tallinn and Tartu, Estonia

Climate
Zone 5, cold and wet winters (down to -30 °C), cool summers (up to 28 °C)

Area
84 m²

Aspect
South- and west-facing garden area

Soil
Loam and clay, improved garden soil, imported garden soil

Wind
Rather sheltered area

Materials
Pinewood decking, seating area, table and raised planter, treated with grey and pink wood stain; galvanized metal structures, mesh and steel wires

Photographer
Merilen Mentaal

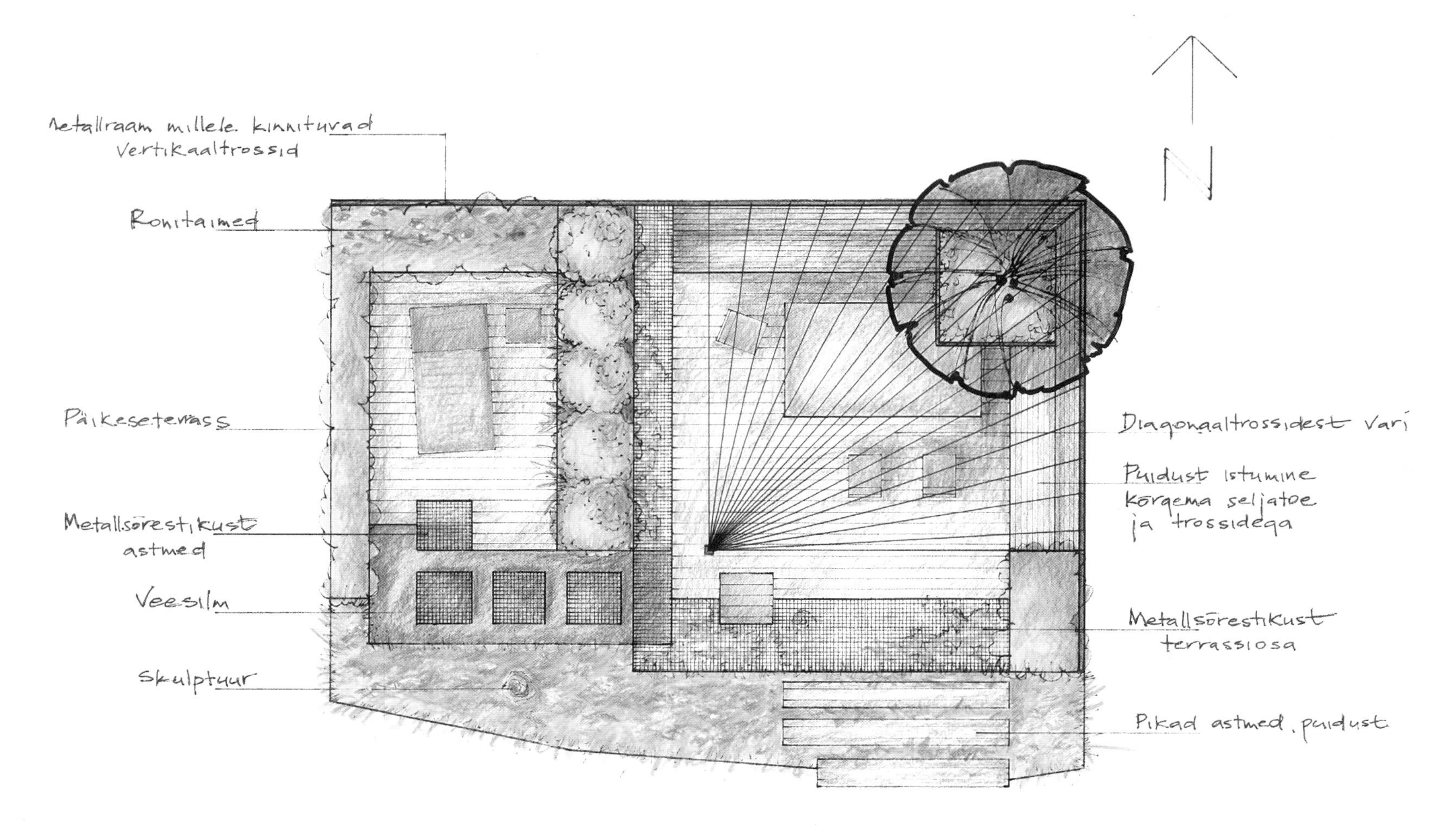

This garden, called "Liam's chic city garden", was created for a flower show lasting for 3 summer months (June-August) to inspire and to show how to use small space that is both comfortable, livable as well as beautiful to be within. The garden has two separate areas – main terrace and sun-terrace, separated by narrow water feature with metal steps crossing over from one terrace to the other. Main terrace has built-in seating, a large table for at least 6 persons to sit comfortably and a large raised planter with Cercidiphyllum japonicum and plenty of herbs to pick for kitchen use (mint, lemon balm, yellow oregano, dark purple basil, variegated sage, rosemary and hybrid blackberry and Tayberry climbing up the wires). Narrow and long strips of wooden steps lead to the main terrace with mat-forming ground cover planting in between them. Smaller sun-terrace is for escaping the crowd and enjoying the sun or afternoon shade from the lounge-chair, surrounded by lush planting and flat, reflecting water.

Cobwebby walls and ceiling was created using steel wires running up and down a metal frame and reaching out like sunrays above the main terrace. These wires will be covered with climbing Lonicera "American Beauty" soon enough, adding further privacy and scent to the garden.

Terraces are separated by larger shrubs – Cercis canadensis "Forest Pansy", Cornus florida "Rubra" and English roses "Young Lycidas" which create green and dark purple contrasts. Colours range from white and pink into dark purple, purple and cerise pink. Three sides of the garden are surrounded by bold perennial planting, mimicking the naturalistic mixtures seen in richer meadows.

Before the end of the festival season, garden found its new home in Tartu, becoming one of the terrace areas of a large town garden. Structures and scale fit perfectly with the existing modern house. Water feature was transformed into planting bed for safety reasons of the couple's young children. Now the narrow stepping stones lead from the garden to house and the view from inside the house greets the eye with bright pink tulips in the spring, whites and blues in summer and autumnal golden tints later in the season.

VAADE EDELAST

VAADE KAGUST

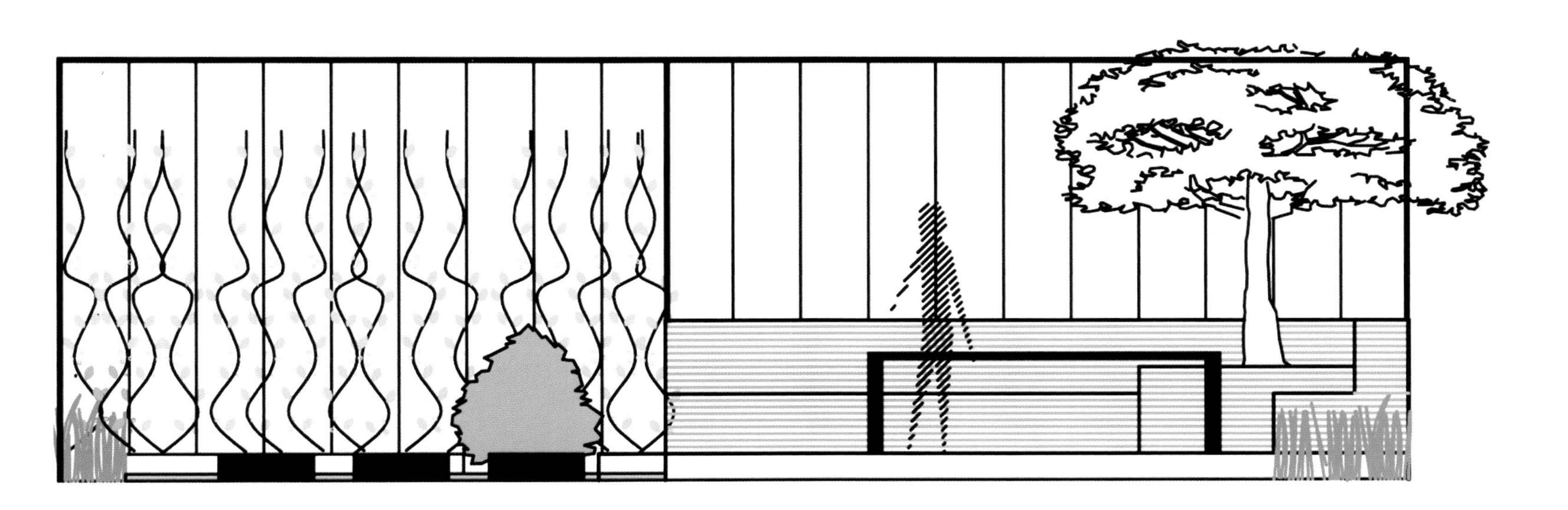

Magical Mediterranean

Landscape Architect
Craig de Necker

Firm
The Friendly Plant (Pty) Ltd

The garden reflects this distinctly Mediterranean atmosphere. This is the kind of garden that you might expect to find somewhere in the Greek Isles, perhaps in Santorini or Mykonos. The rustic cladding and woodwork add to the theme, which is punctuated by the use of greys, whites and blues in the garden.

The designers selected composite decking boards in order to fabricate the fence as well as to construct the rustic awning over the patio. This provides a clean, contemporary look, and is environmentally-friendly and maintenance-free.

Planting makes use of trees such as Olea europaea and Buddleja saligna, shrubs and perennials with blue or white flowers such as Freylinia tropica, Agapanthus and Rosemarinus officianalis to add a true flavor of the Mediterranean.

Water and fire are two of the elements that one can watch for hours on end. Their movement is so magical and delicate, yet they both hold enormous power. The fire provides warmth and atmosphere to the area and the water feature provides a relaxing background sound.

The unique water feature is integrated with a fire pit and is cantilevered over the pond below. Water gently flows from the fire pit, over a rim into the water underneath. The designers used glass mosaic in order to bring in some vibrant colour and to include more of the colours of the Mediterranean.

The centerpiece behind the seating area is a commissioned artwork, depicting the summer sky. The blue and white tones in the painting work well to bring together the other colours found in the garden.

This garden is a place for contemplation and relaxation, allowing one the opportunity to enjoy watching the world go by. In our fast-paced world where email and phone calls can find you almost anywhere, this is a garden that makes it easier to simply 'switch-off' and relax.

SF Residence

Landscape Architect
Ron Lutsko Jr.

Firm
Lutsko
Associates landscape

Structural Engineer
Kris Johnson, Randy Braun, GFDS Engineers

General Contractor
Plath and Company, Inc.

Bronze Wall w/ Water
Eric Powell

Steel/Glass Panels and Wall Frames
Conceptual Metal Works

Plaster
Tony Olea

Mason
Mike Petty

Landscape Contractor
Dan Fix Landscape Construction

Photographer
Marion Brenner Photography

This urban garden in the Pacific Heights district of San Francisco surrounds a single family home. The house was designed in 1990 by the noted Bay Area architectural firm EHDD under the tutelage of the late Joseph Esherick. When viewed from the home's two balconies overlooking the site, the garden is a graphic composition of space, materials and planting. From this perspective the distant San Francisco Bay and Alcatraz Island slip into the view of the city spread out below the property. The garden is designed to carry the sense of the city's urban fabric into the site. From within, the garden is intimately experienced as a sequence of outdoor rooms, their spaces and design features sequentially concealed and revealed as one passes through each threshold in the composition.

The garden is divided into a series of three enclosed garden rooms, each rigorously defined by architectural and/or planted edges. The variety of edge treatments – translucent glass, hand-troweled plaster, a curved bronze wall, and planting – explore the relationship between viewer and the adjacent off-site conditions. The walls between rooms are composed of clipped Prunus caroliniana hedges, fit within the structure of steel frames. Thresholds between the spaces create a sense of mystery and discovery as one moves through the garden.

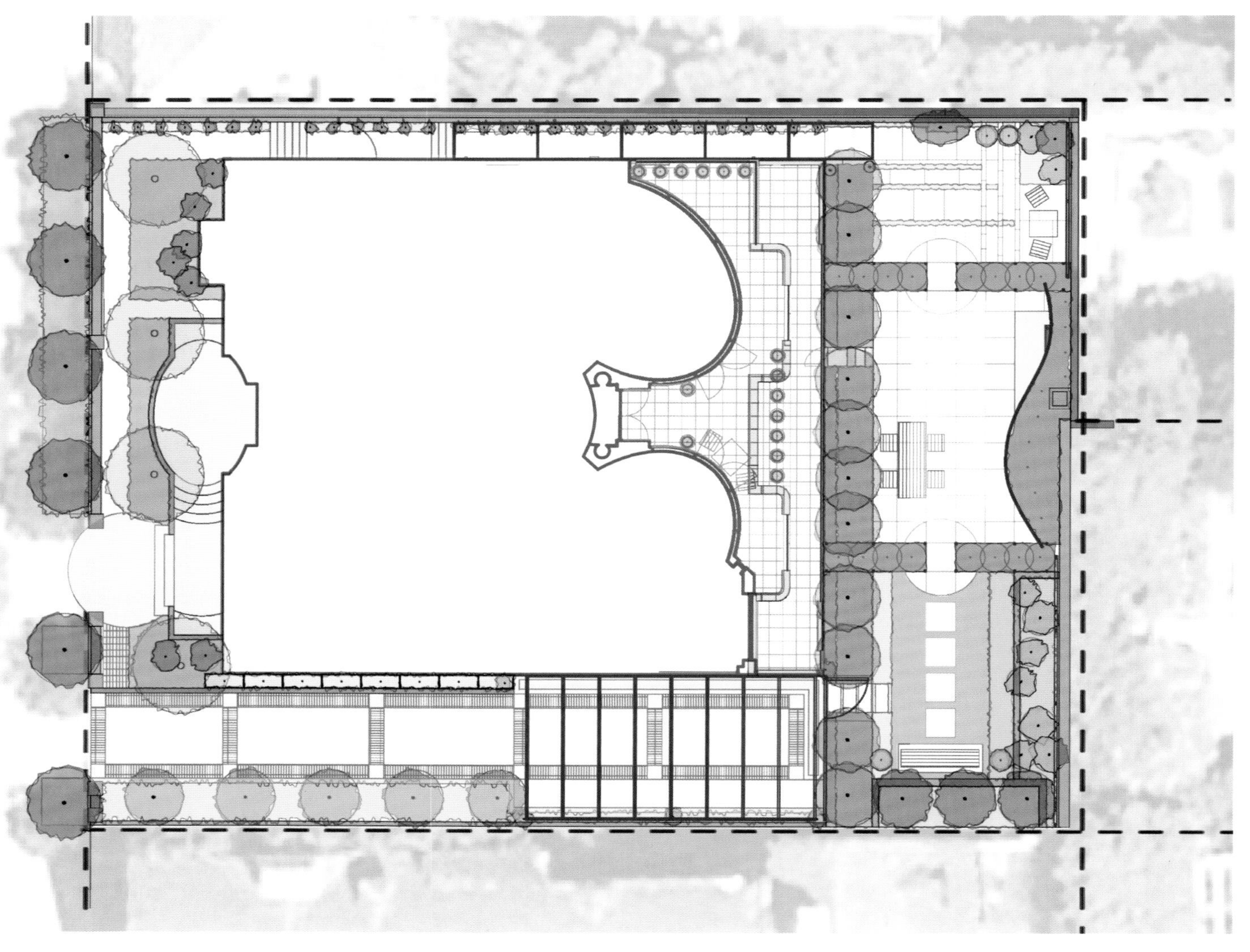

The Australian Garden Presented by Royal Botanic Gardens Melbourne at the Chelsea Flower Show 2011

Landscape Architect
Jim Fogarty

Firm
Jim Fogarty Design Pty Ltd

Location
London, UK

Area
220 m²

Photographer
Jim Fogarty, Jay Watson
Courtesy of Royal Botanic Gardens Melbourne

In telling the story of the journey of water, the front of the garden starts with the arid outback of Australia. The front entrance path is made from impregnated tumbled quarry pebbles to symbolise a dry riverbed. The water flows along the water feature and when it reaches the coastal part of the garden it cascades back down into the ground aquifers.

The sculpture is laser cut and has been coated in a fibreglass resin for strength. The sculpture is 3-dimensional so it casts shadows by the sun. To get the right effect, the designers lightly dusted actual red sand from the Australian Garden over the wet paint.

Each band of the waterhole represents a different sedimentary layer symbolic of an outback waterhole. A waterhole in the outback is a wet soakage pit that dries out during extended periods of dry.

The 'Salt' sculpture is a remodel of the Ephemeral Lake sculpture at the Australian Garden in Cranbourne. The shapes of the salt are inspired by satellite images of outback saltpan areas such as you would find in the Simpson Desert in outback Australia.

Green grass plays an important role in combating dust around cattle station homesteads in the Outback. Grass provides a cooler microclimate in areas where day time temperatures can reach over 40 degrees. Having grass in this garden also symbolises the urbanisation of Australia. The shape of the lawn depicts the farmed green valleys found across the Great Dividing Range which separates inland Australia with the East coast.

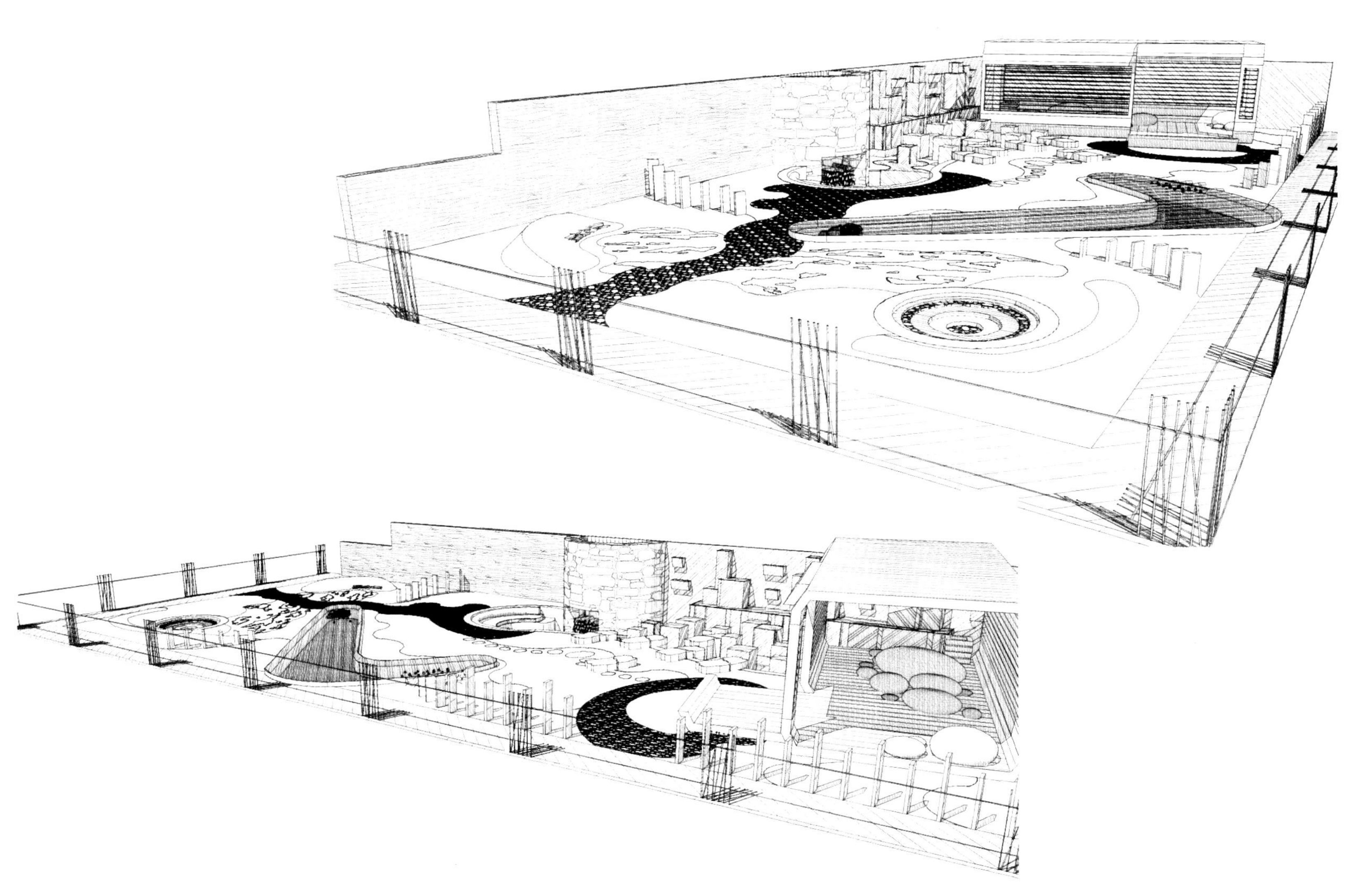

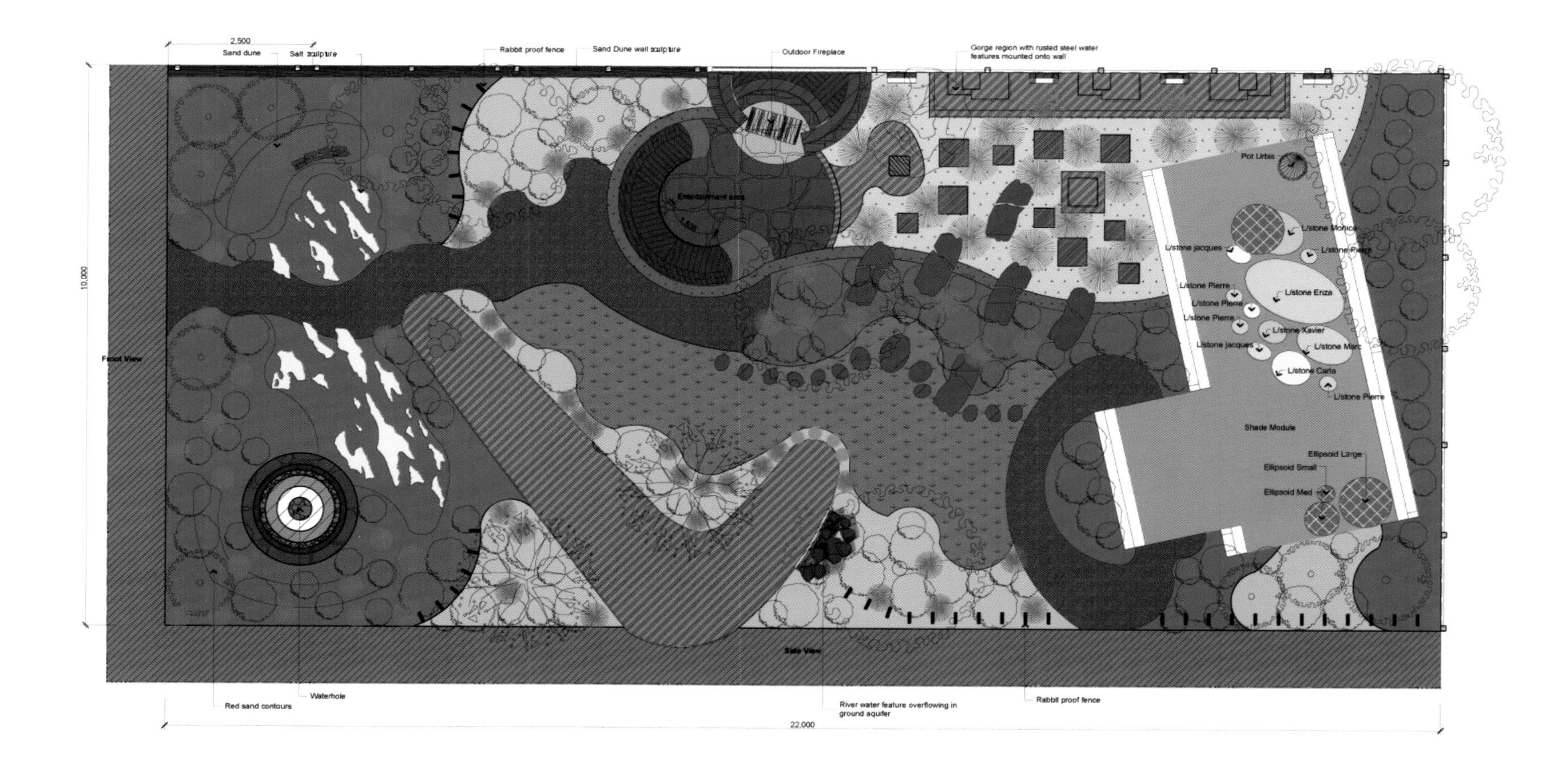
2,500
Sand dune
Salt sculpture
Rabbit proof fence
Sand Dune wall sculpture
Outdoor Fireplace
Gorge region with rusted steel water features mounted onto wall
Entertainment area
10,000
Front View
Pot Urbis
L/stone Monica
L/stone jacques
L/stone Pierre
L/stone Enza
L/stone Pierre
L/stone Pierre
L/stone Pierre
L/stone Xavier
L/stone jacques
L/stone Marc
L/stone Carla
L/stone Pierre
Shade Module
Ellipsoid Large
Ellipsoid Small
Ellipsoid Med
Side View
Red sand contours
Waterhole
River water feature overflowing in ground aquifer
Rabbit proof fence
22,000

22,000
2,400
6,900
2,828
+2,600
2 Level 1
±0
1 Ground
-750
0 Foundation
9,872
Sand Dune Wall sculpture
10,000
600
2,000

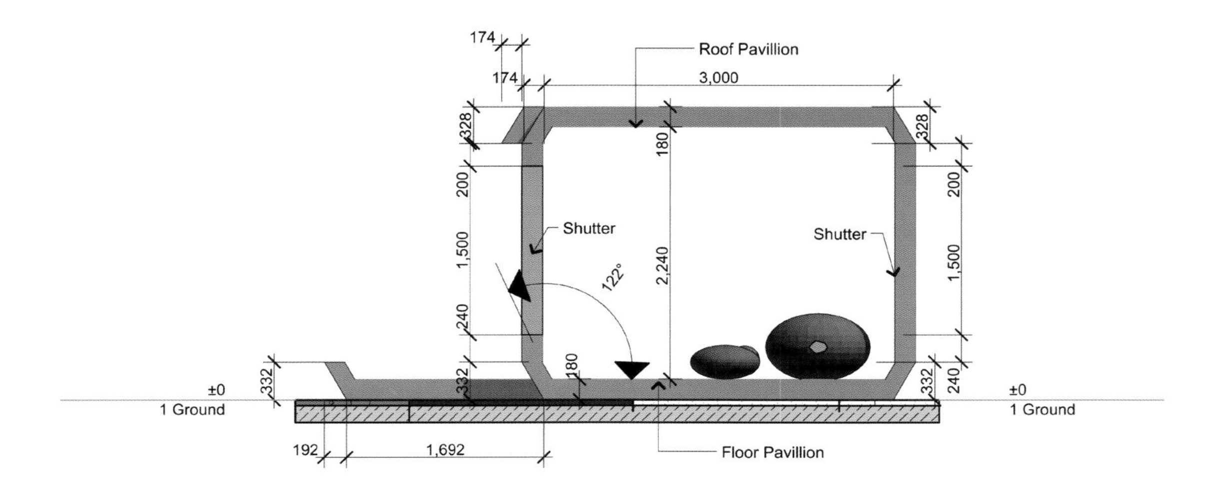
174
174
Roof Pavillion
3,000
328
180
328
200
200
Shutter
Shutter
1,500
2,240
1,500
122°
240
180
332
332
332
240
±0
1 Ground
±0
1 Ground
192
1,692
Floor Pavillion

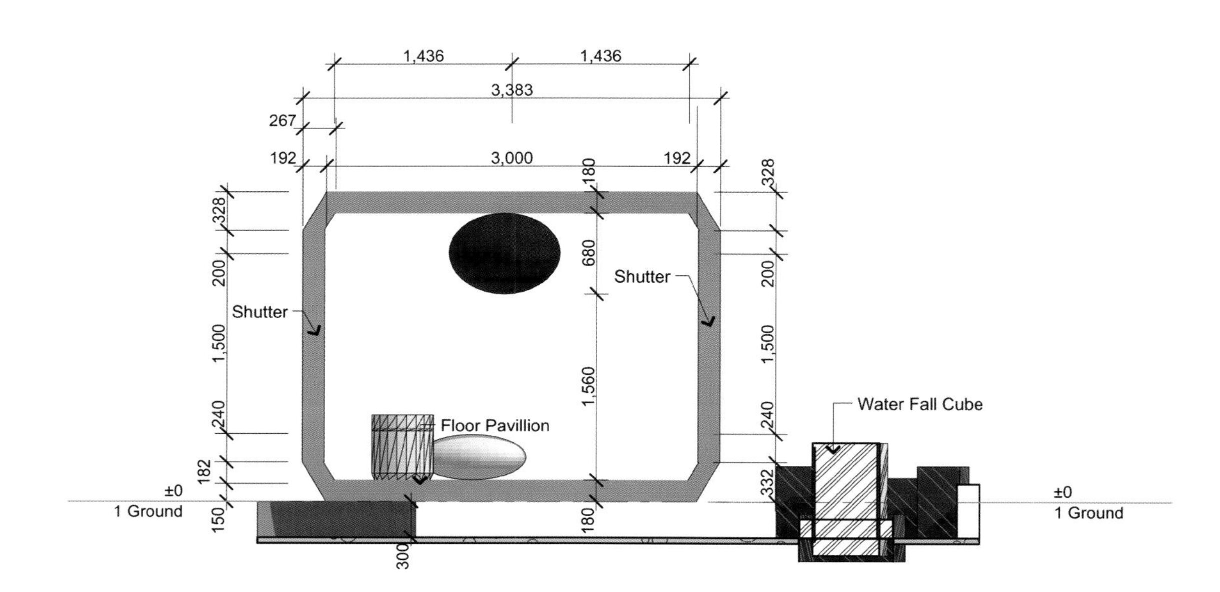
1,436
1,436
3,383
267
192
3,000
192
180
328
328
680
200
200
Shutter
Shutter
1,500
1,560
1,500
240
240
Water Fall Cube
Floor Pavillion
182
332
±0
1 Ground
150
180
300
±0
1 Ground

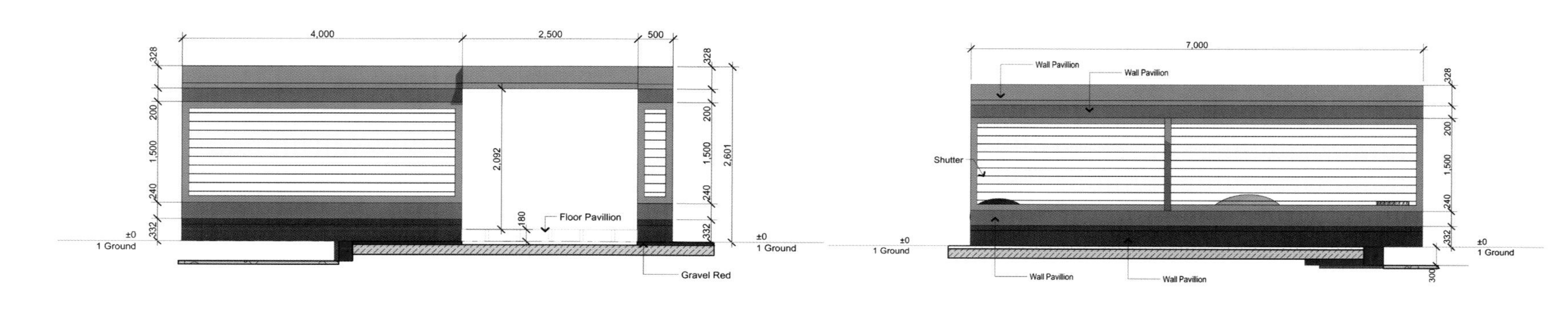
4,000
2,500
500
328
328
200
200
1,500
2,092
1,500
2,601
240
240
180
Floor Pavillion
332
332
±0
1 Ground
±0
1 Ground
Gravel Red
7,000
Wall Pavillion
Wall Pavillion
328
Shutter
200
1,500
240
332
±0
1 Ground
±0
1 Ground
Wall Pavillion
Wall Pavillion
300

Riseley

Landscape Architect
Scott Brown Landscape Design

Spa
Endless swim spa (acrylic, with ceramic tiles to waterline)

Paving
Concrete pavers, on concrete base
Driveway, saw-cut concrete. 'In-situ'

Decorative Screen Panels
Rendered compressed cement sheeting on freestanding backing frames (steel/timber)

Front Fence
Rendered masonry, wrought iron panel inserts and gates

Planter-box Walls and Steps
Rendered brick (masonry)

Water Feature
Concrete shell with acrylic base pond (Rietmans)

Outdoor Lighting
Low voltage (Halogen), Hunza (N.Z.), Megabay (N.Z.), Lumascape (AUST) and Castlight

Plantings
Boundary hedges- Lilly-pilly varieties (Syzygium Australe), (Acmena Smithii)
Small border hedging- Dutch Box (Buxus sempervirens suffruticosa)
Feature Trees (Front)- Designer Indian Bean Tree (Catalpa bignonioides 'Nana')(Flemmings)
Camellia Sasanqua, Azaleas ('Alba Magnifica'), Hebe ('Wiri Spears') and box-leaved privet (Ligustrum Undulatum)

Location
Melbourne, Victoria, Australia

When you enter the front garden you will be immediately struck by the elegant formality of the grounds, both of which tie in with the architecture of the home.

One of the most critical properties of both the front and rear gardens is they are quite drought-resistant. Considering the current water predicament, this is obviously a much needed prerequisite when designing a new garden, but it's not a feature usually associated with such a formal garden design.

The use of boundary hedges to screen the neighbouring homes and gardens, as well as hide the boundary fences, has given the grounds a lush, timeless appeal. By using native Lilly-pilly hybrids – Syzygium australe in the front and a select form of Acmena smithii in the back – these hedges have contributed to the enduring, classical look of the property, but without the need to introduce into the mix plants with high watering needs.

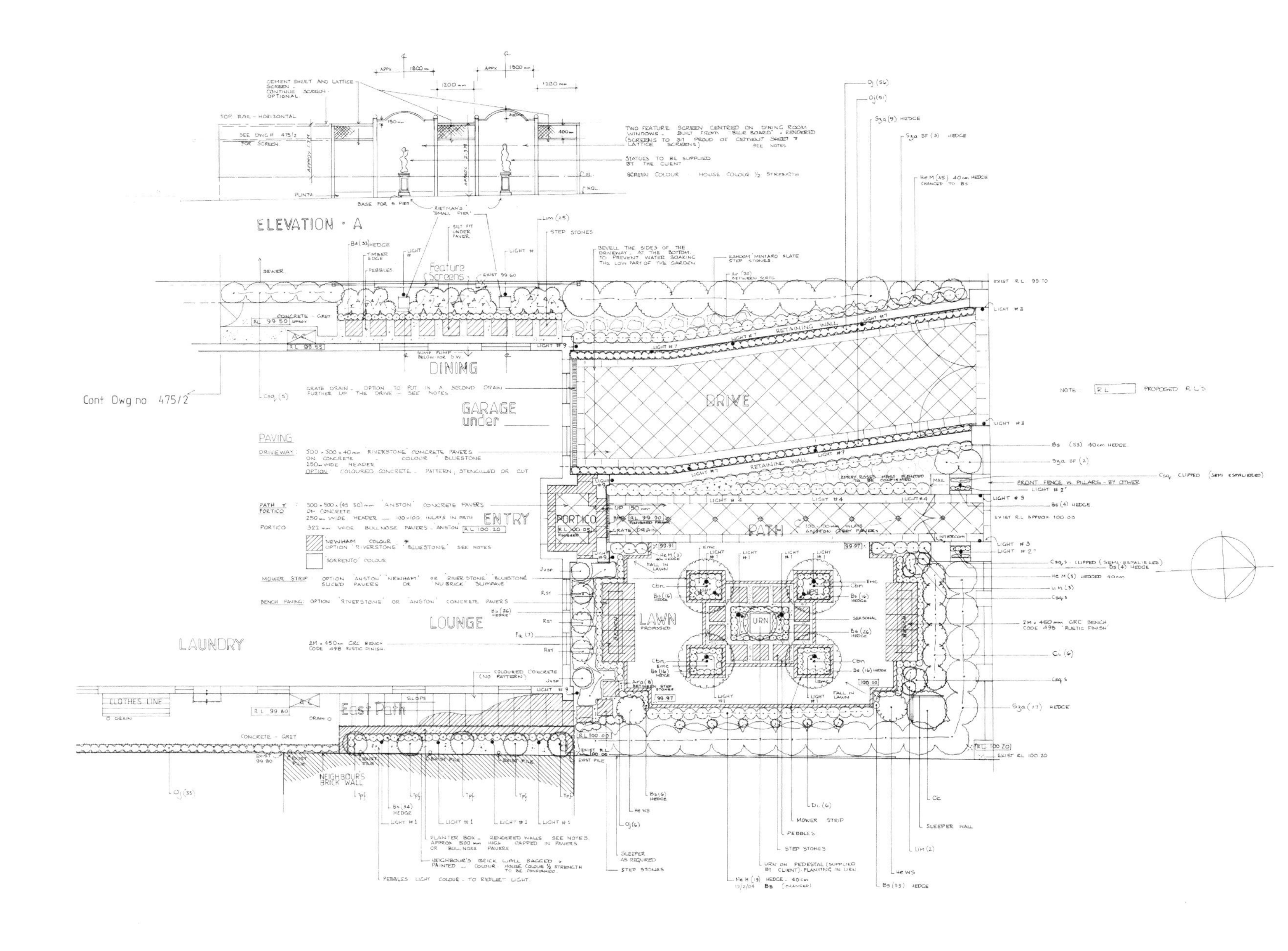

The front specimen lawn faces north, which means the entry courtyard can get quite warm on a sunny day. With this in mind, the designers used fine leafed Santa Ana Couch. This type of turf can develop a high degree of drought resistance and will repair itself should any sections die off during particularly long dry spells. The other planting consists of repeated hedges, adding to the structural integrity of the garden. These include Dutch box (Buxus sempervirens 'Suffruticosa') and rows of Liriope 'Royal Purple', a very drought-resistant plant. Hebe 'Wiri Spears' (a New Zealand native), standard roses and four matching specimens of 'designer' Indian bean tree (Catalpa bignonioides) complete the front courtyard picture. The Indian bean trees surround a large Grecian urn in the middle of the lawn, which gives this small entry courtyard a focal point.

The garage is beneath the house and contains substantial tanks for water run-off and seepage collection. This water storage, combined with the hardy nature of most of the pants, ensure the garden is able to deal with Melbourne's ongoing water restrictions.

To complete the look in the front garden, the designers used sand-blasted concrete pavers (from Anston) and appropriate furniture in the form of a garden seat.

The back garden contributes a wonderful outlook that can be enjoyed from inside the house all year round, while also catering for the family's every outdoor entertaining requirement. The swim spa, for example, not only provides an intimate experience for one to eight people; it's also a great exercise pool. For maximum convenience, the spa is positioned close to the toilet/bathroom built in the corner of the rear courtyard.

The dining area is surrounded by the house, a garden area and a water feature-fountain, and is situated to one side of the courtyard. This provides a more intimate dining experience and helps to separate diners from the noise of the spa. At night, lighting ensures this area takes on a whole new dimension.

The patio attached to the house is partially enclosed with a hedge of Box Leaved Privet (Ligustrum undulatum), which forms a living balustrade. This continues the theme of partially dividing the rear courtyard into smaller sub-rooms, the result being more interest and more intimacy.

The western side of the home has a hedge of Camellia Sasanqua and there are statuettes positioned outside the window of the formal dining and lounge rooms. The lighting and the rendered screen built to form the backdrop provide a stunning outlook both day and night from these rooms of the house.

The garden's necessary evil – the clothes line – is on the eastern side of the house. Because there are no ground-floor windows here, this area has been designated the utility zone of the garden.

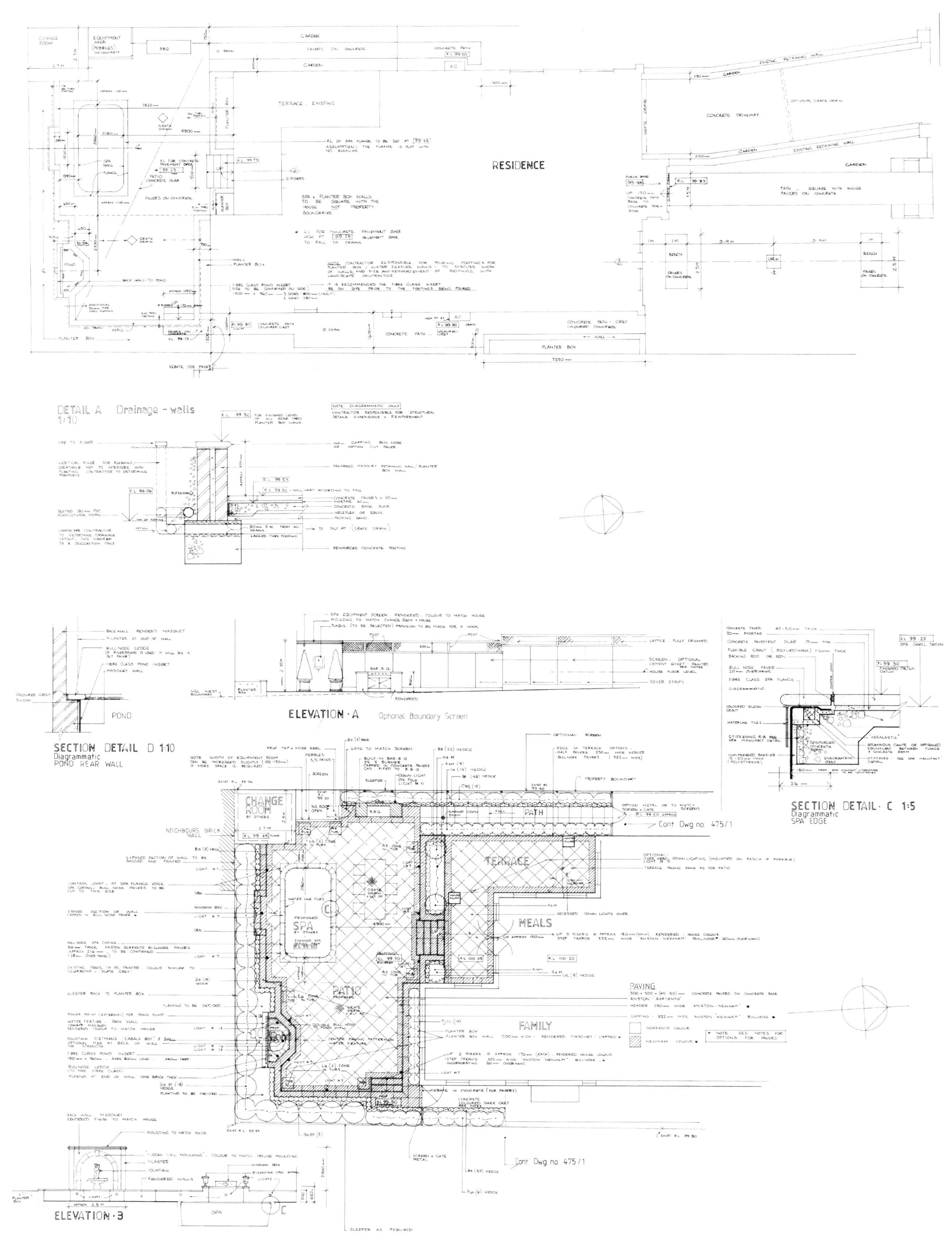

RESIDENCE
GARDEN
TERRACE . EXISTING
CONCRETE DRIVEWAY
DETAIL A Drainage - walls
1:10
POND
SECTION DETAIL D 1:10
Diagrammatic
POND REAR WALL
ELEVATION · A
Optional Boundary Screen
SECTION DETAIL · C 1:5
Diagrammatic
SPA EDGE
CHANGE ROOM
PATH
TERRACE
MEALS
PATIO
SPA
FAMILY
PAVING
Cont Dwg no 475/1
ELEVATION · 3

Leafy Entertainer

Landscape Architect
Scott Leung & Myles Broad

Firm
Eckersley Garden
Architecture

Well facilitated multi-purpose spaces, interlaced and layered with selected plantings are appealing. Spaces where the plants envelop and entwine create the experience of living in the garden. The main concept with the design of this Melbourne project was to do just that. If the success of a space is how often it gets used, then this design is very successful as it gets a lot of use.

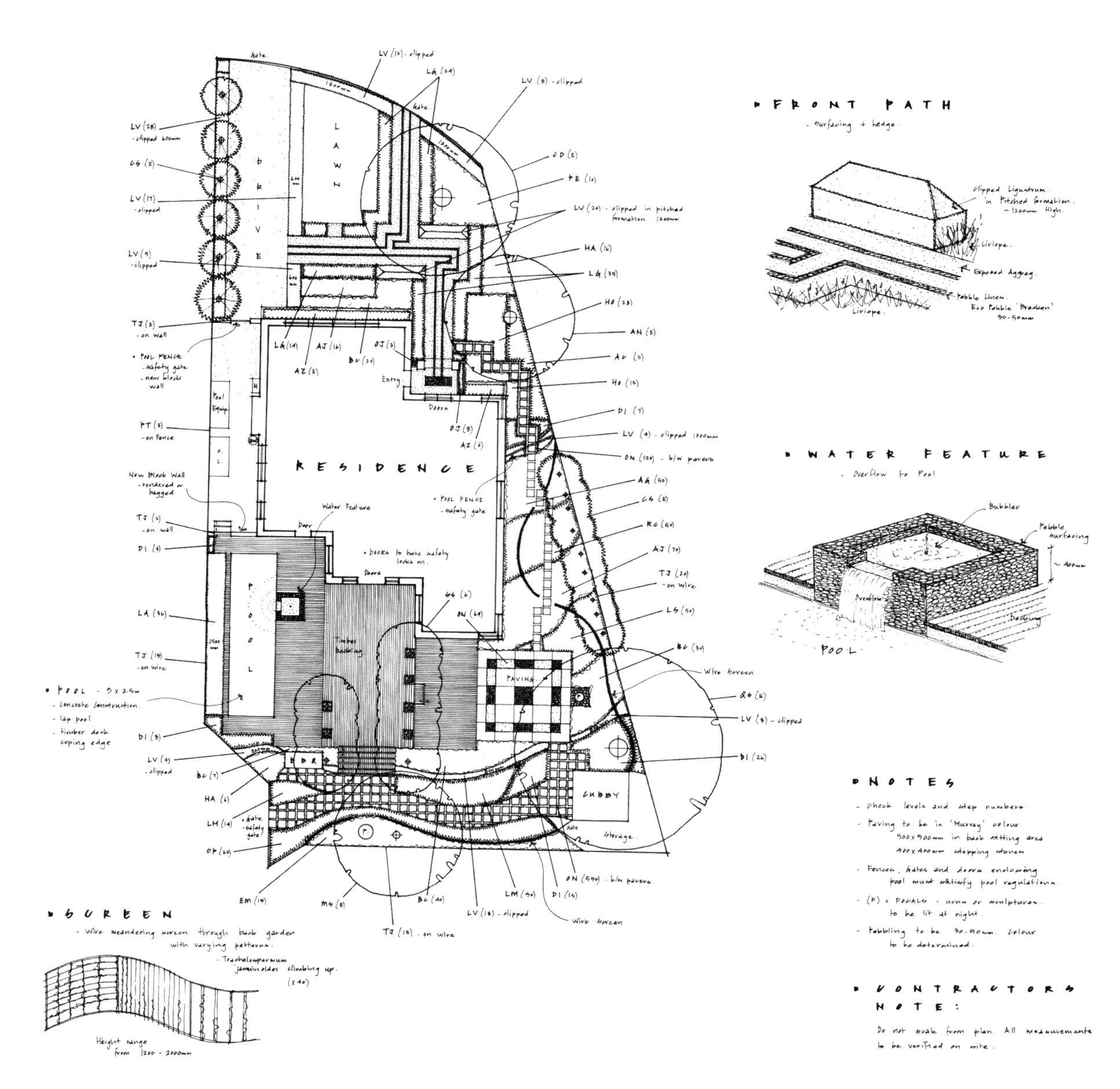

C   emporary City Gar

Landscape Architect
Cherry Mills

Firm
Cherry Mills Garden Design

Location
Guildford, England

This town centre Regency property had been completely renovated by the clients in an elegant, contemporary style. It was now time to sweep away the dilapidated and overgrown garden, front and back. The clients, both busy professionals, wanted a cool, slick garden with minimal maintenance requirement. Initially the designers had to replace a crumbling brick boundary wall and remove a number of mature trees which shaded the garden for much of the day. As the house is Listed and in a Conservation Area, all the relevant permissions were sought and obtained from the Local Authority. The sloping site prompted the division of the garden into three outdoor 'rooms' at different levels. Water is a key feature of the garden and three separate fountains were designed and installed. Pale Portland stone was chosen for the paving and a blue/purple and white color scheme for the planting, including Salvias, Agapanthus and Verbena Bonariensis. Two rows of pleached Hornbeams and horizontal slat trellising were installed to give additional screening along the boundaries. A token area of artificial 'turf' provides a decorative frame around the dining area. A comprehensive lighting scheme brings the whole garden to life at night and adds a new dimension to the garden.

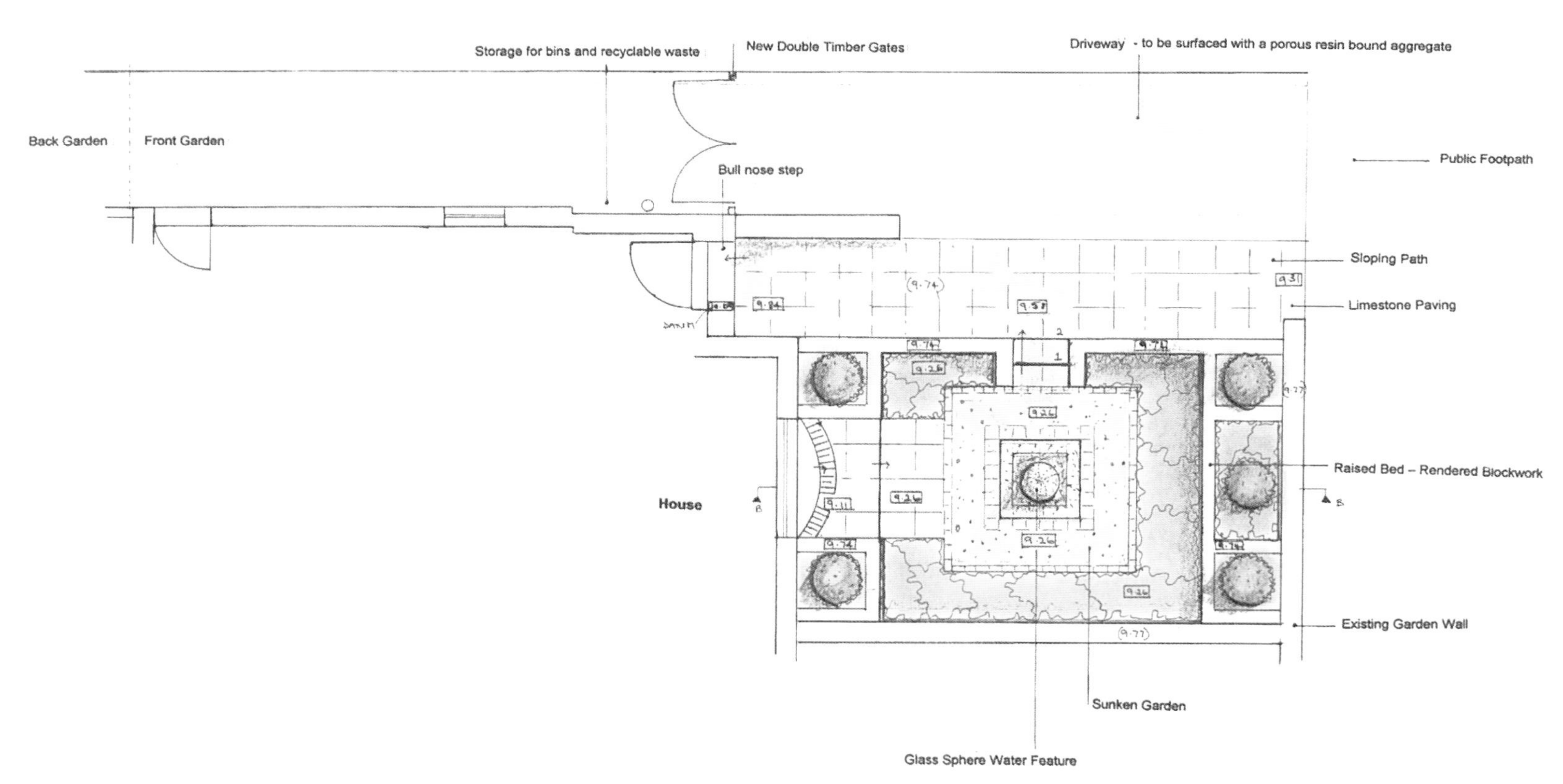

Front Garden

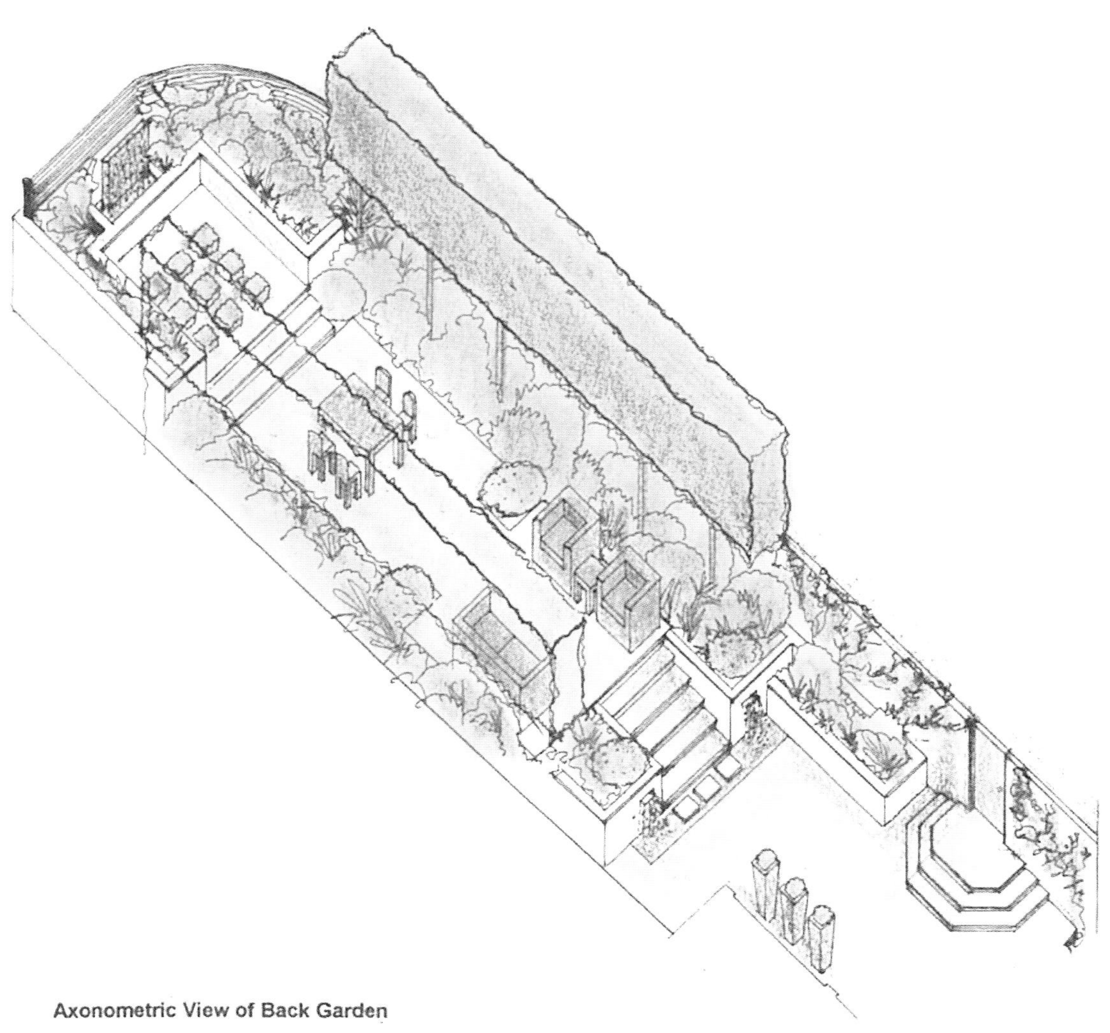

Axonometric View of Back Garden

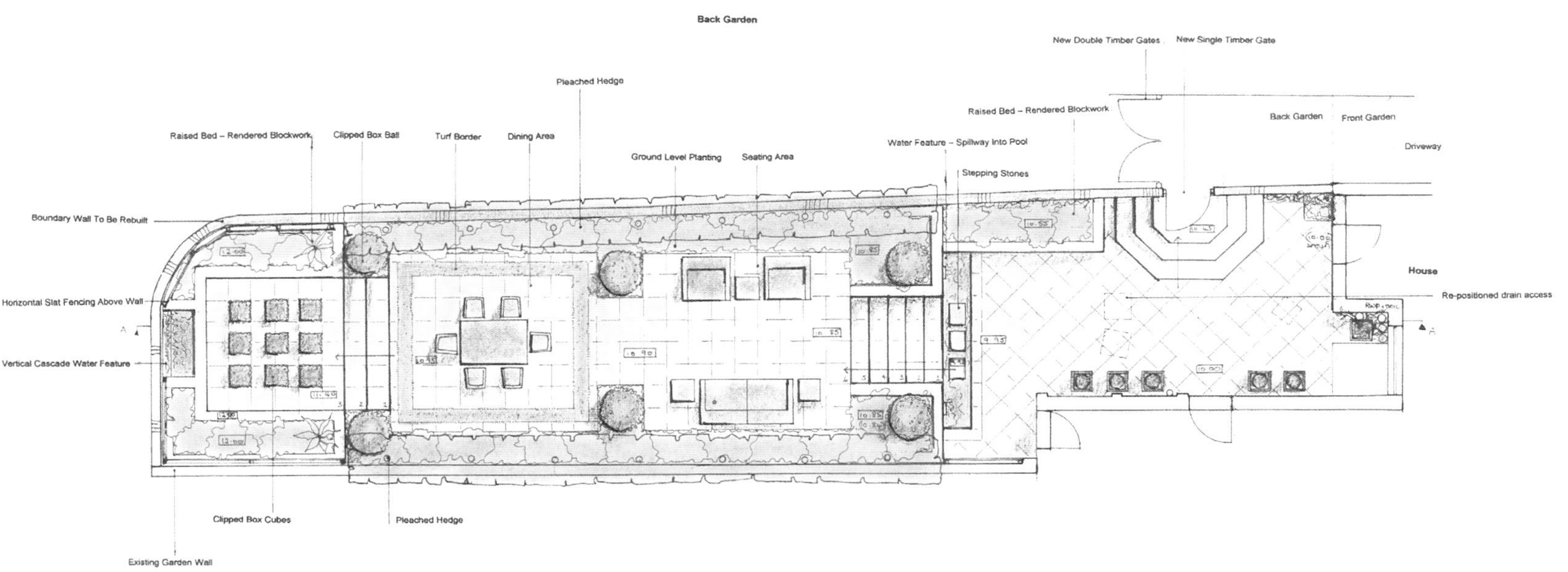

Back Garden
New Double Timber Gates
New Single Timber Gate
Pleached Hedge
Raised Bed – Rendered Blockwork
Raised Bed – Rendered Blockwork
Clipped Box Ball
Turf Border
Dining Area
Ground Level Planting
Seating Area
Water Feature – Spillway Into Pool
Stepping Stones
Back Garden
Front Garden
Driveway
Boundary Wall To Be Rebuilt
Horizontal Slat Fencing Above Wall
Vertical Cascade Water Feature
House
Re-positioned drain access
Clipped Box Cubes
Pleached Hedge
Existing Garden Wall

Country Garden Radlett

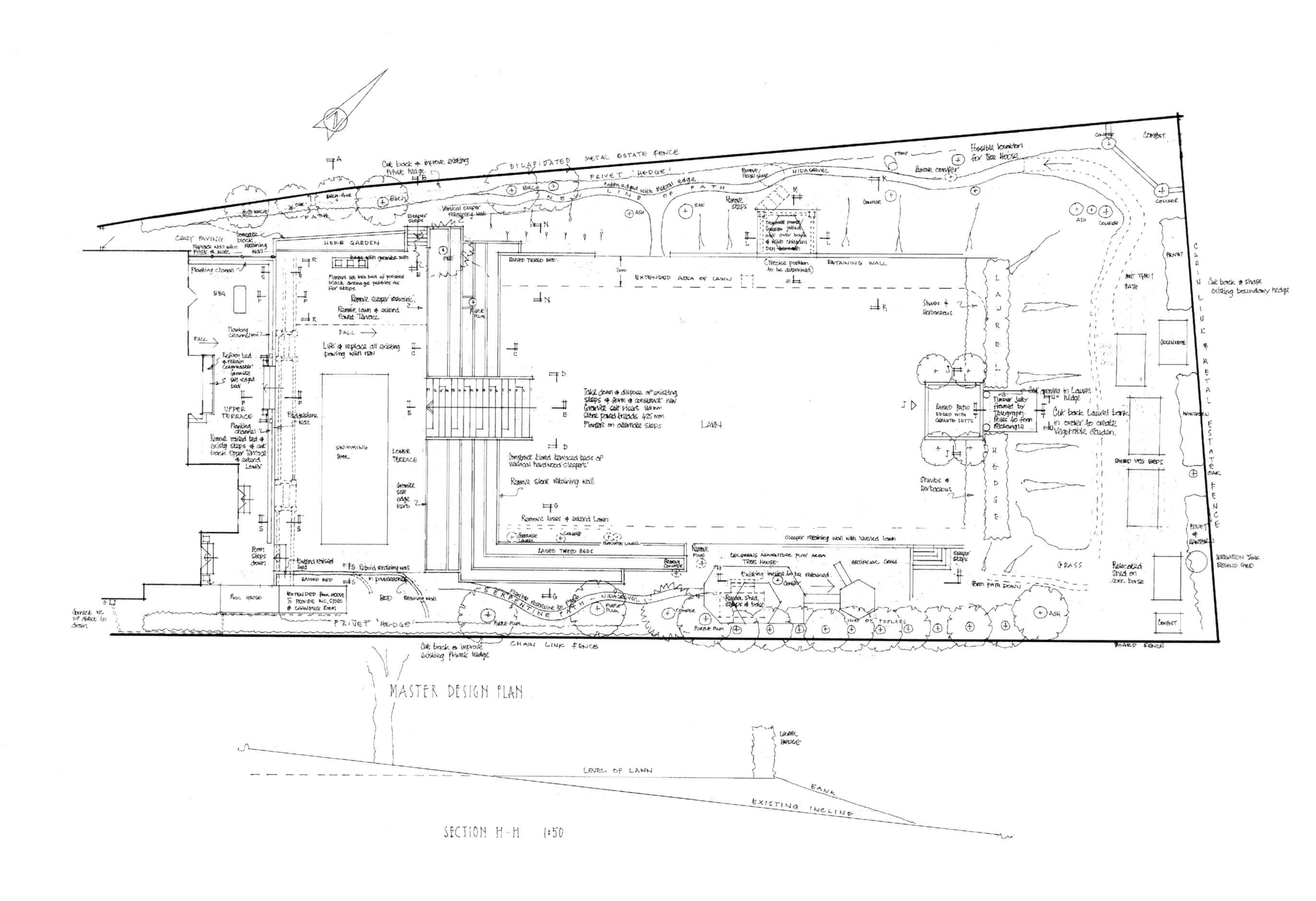

Landscape Architect
John Nash

Firm
John Nash Associates

Location
Radlett, Hertfordshire, England

Overgrown, dense vegetation and weed filled; uneven and broken concrete paving and uninspiring tired flower beds, this was the original garden and did little to provide pleasure, despite the presence of the Swimming Pool. The owners, Phil and Emma, were conscious that their garden did not realise its full potential and they invited the designer to see what he could do to improve it. It had lovely sweeping views down to a valley with hills beyond. They wanted to increase the width and depth of the lawn, the bottom of which was intersected by a high Laurel hedge which broke up the view and foreshortened the depth of the garden.

The steep bank leading down from the lower terrace to the lawn was broken up with a series of tiered terraces, held in place by vertical oak timbers and the dangerous narrow and uneven central section steps replaced with inviting, much wider steps, paved the same as the terrace with alternate treads stopping short of full width to enable Box ball planters to be built in.

The lawn area was increased in depth and width by cutting the steep banks on each side with a continuation of the oak retaining walls. The tiered terraces were planted with Box and Euonymus Silver Queen. Phil and Emma didn't want the distant view to be obscured from the terrace so the planting between the lower terrace and top of the bank needed to be carefully considered. The designer proposed Stipa tenuissima, a fairly low growing grass but which provides sensual movement as it catches a whisper of wind and in the spring, colour is introduced by Tulips and Alliums and later on in the summer with Cosmos.

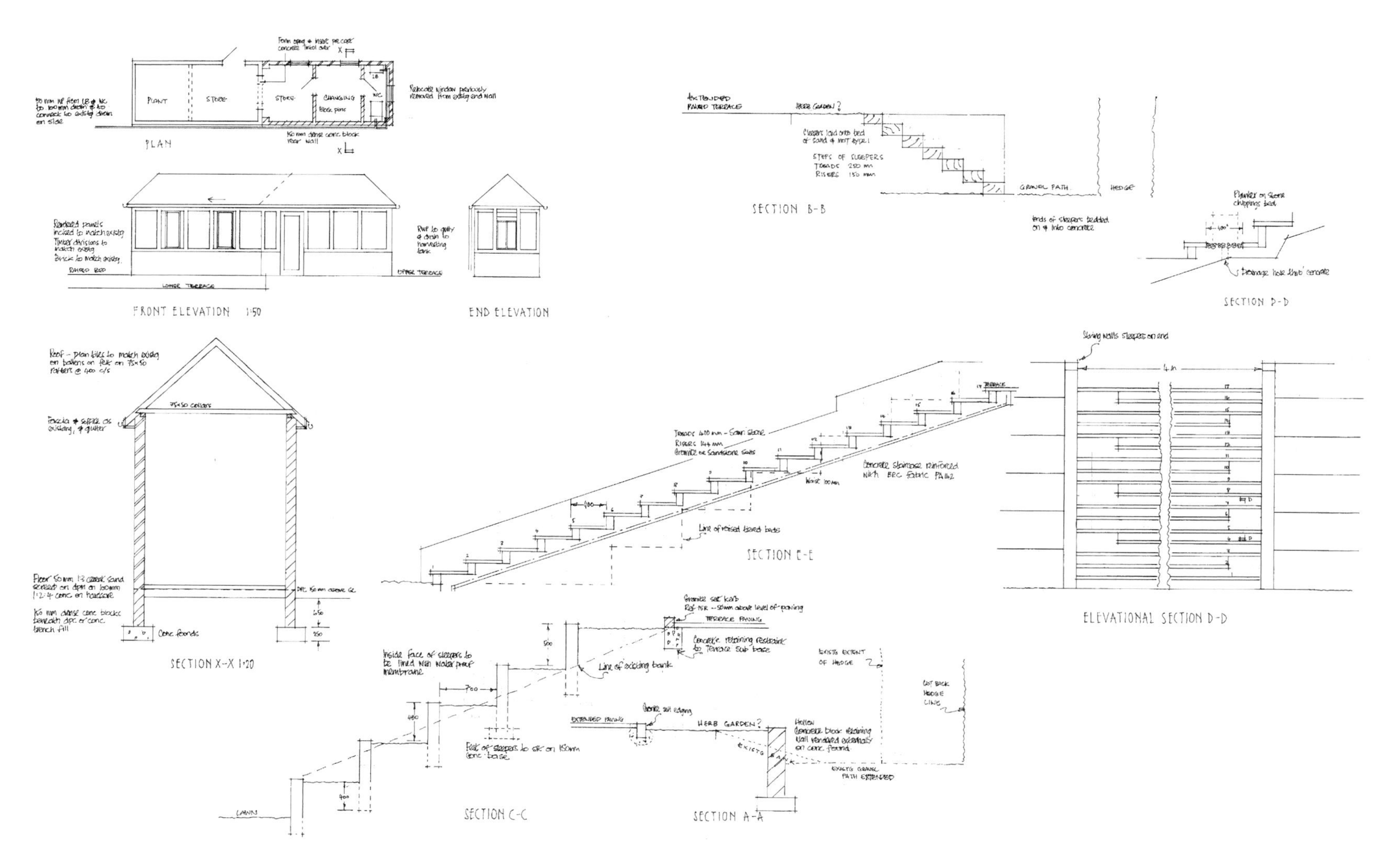
PLAN
FRONT ELEVATION 1:50
END ELEVATION
SECTION B-B
SECTION D-D
SECTION X-X 1:20
SECTION E-E
ELEVATIONAL SECTION D-D
SECTION C-C
SECTION A-A

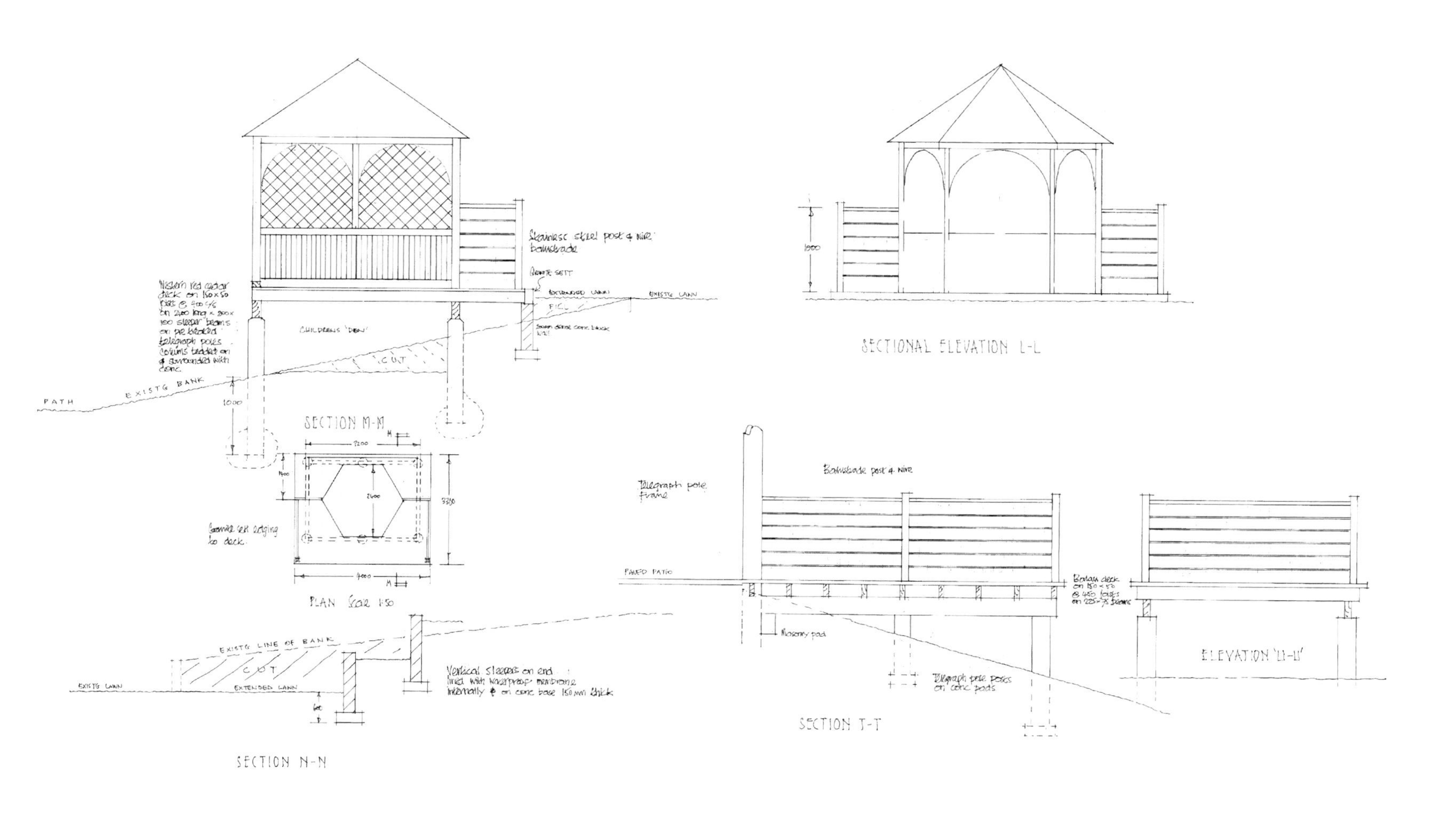
SECTION M-M
SECTIONAL ELEVATION L-L
PLAN Scale 1:50
SECTION N-N
SECTION T-T
ELEVATION 'U-U'
PATH
EXISTG BANK
CUT
CHILDRENS 'DEN'
EXISTG LINE OF BANK
EXISTG LAWN
EXTENDED LAWN
PAVED PATIO
Telegraph pole frame
Balustrade post & wire
Masonry pad
Telegraph pole posts on conc pads
1000
1200
4000

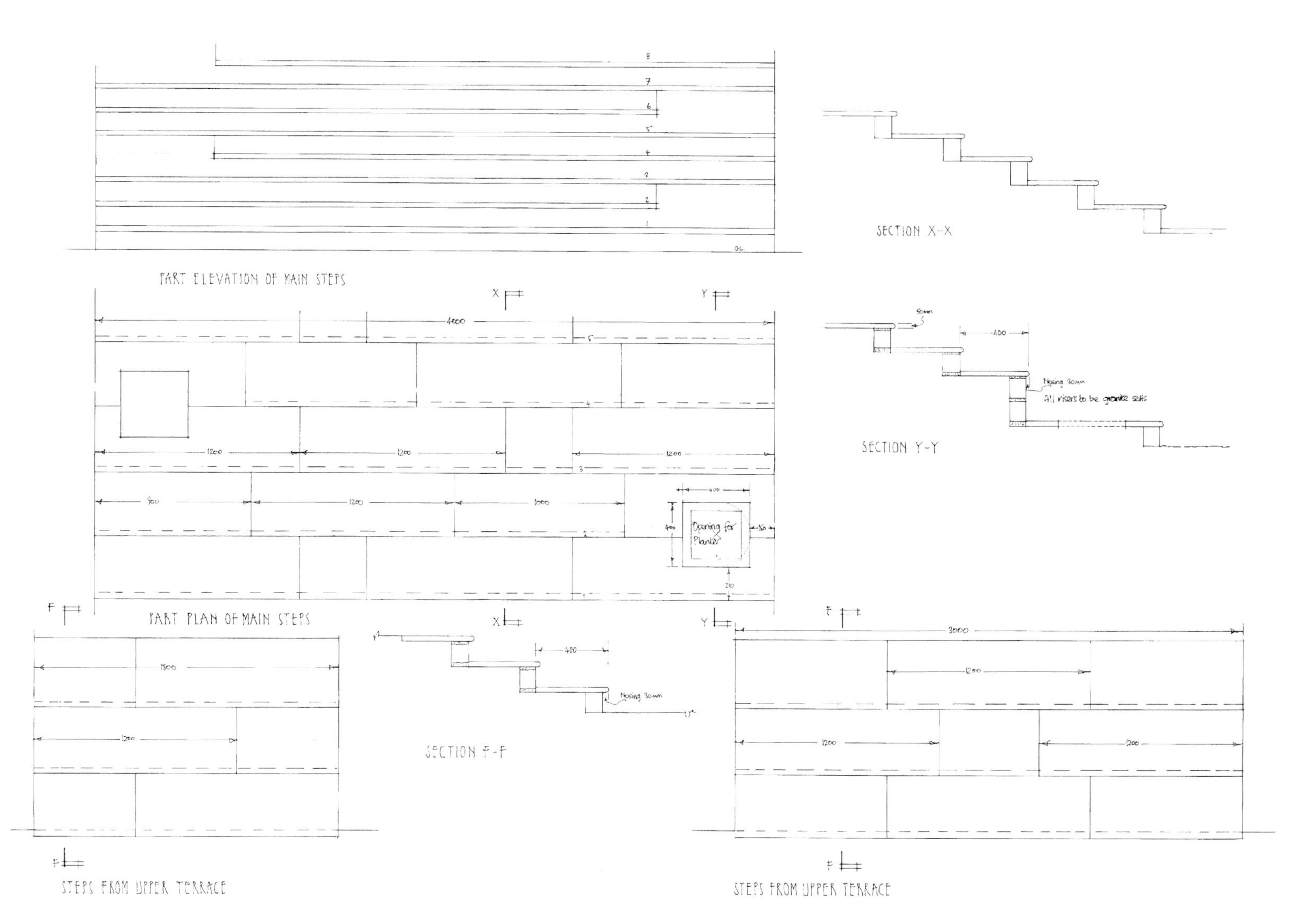
PART ELEVATION OF MAIN STEPS
SECTION X-X
PART PLAN OF MAIN STEPS
SECTION Y-Y
SECTION F-F
STEPS FROM UPPER TERRACE
STEPS FROM UPPER TERRACE
Nosing 30mm
All risers to be granite setts
Opening for Planter
4000
1200
900
1000
400
2000

Donnellan

Landscape Architect
Scott Brown Landscape Design

Front Driveway (paving)
Exposed stone aggregate concrete, with formed concrete borders to match front path and portico pavers

Front Path and Portico Paving
Concrete pavers (Riverstone, 500mmx500mm)

Swimming Pool
Reinforced concrete(sprayed)

Shell Internal Finish
Ceramic tiles

Pool Coping and Main Pavement Edging
Concrete pavers

Main Pavement
Exposed stone aggregate (concrete matrix)

Pool Fencing
Semi-frameless glass with powder-coated aluminium posts powder-coated aluminium frame (lawn end)

Dining Setting
GRC table top, with 'Roman' leg bases, and powder-coated aluminium chairs

Planting
Central feature tree- Weeping Mulberry (Morus alba 'pendula')
Twin feature trees- 'Designer'Maple (Acer platanoides'Globosum') (Flemings)
Low border hedging- honey suckle (Lonicera nitida)

Other Plantings
Camellia sasanqua 'Set-su-gekka', 'ice-berg' standard roses, pencil pines (Juniperus virginiana 'Skyrocket'), French Lavender (Lavandula Dentata), seaside daisy (Erigeron mucronatus) and Hebe 'Inspiration'
Twin water fountains (front)- Concrete (Paddington Pots)

Construction Period
12 months

Location
Melbourne, Victoria, Australia

The design brief may have been straightforward – relate the structured elegance of the residence to its wild, natural bushland setting – but reconciling the inherent contradictions this posed required a great deal of thought.

Because the front yard was relatively level, unlike the back garden with its steep natural fall, the designers were able to implement a rather grand, formal design which imbues the house with a greater 'sense of place' when viewed from the street. At the back of the house, however, they needed to resolve the challenges presented by the precipitous fall. This was achieved by creating a 'floating' pool and an adjacent entertainment area, both supported by pillars.

The back garden has been designed to function as a series of structured outdoor 'rooms'. The sheer, three-four metre drop-off edge of the pool, along with the fenced perimeter of the outdoor dining and lawn/garden areas, separate the living and entertainment area at the rear of the home from the untamed natural landscape below. This lower part of the block merges with the surrounding countryside, connecting the garden to a larger environment dominated by rolling hills, rocky gullies and clump plantings of manna gums.

To create such an easily accessible and functional living area required significant engineering expertise. To imbue the space with such a sense of luxury required just the right use of materials, including the exposed aggregate pavement, the grey-black pool coping and pool fence posts, the large concrete planters and the wrought iron dining chairs – all of which work together to create a unified look.

In the front yard, access to the five-car garage is via an expansive semi-circular driveway where the customised exposed aggregate surface tones down what might otherwise have been an overly dominant feature. The colours of the stones used in the driveway match the colours expressed in the exterior finishes of the residence – namely charcoal, grey, cream and white. As buff sandstone-coloured concrete pavers were used for the front porch and path, they were also used as a decorative border along the driveway thereby visually linking the key hard landscape elements.

As a further complement to the quiet elegance of the home, the floral colour palette in the front garden is restricted to white, cream, mauve and maroon. Foliage colours vary from dark green to mint green, and from grey-green to blue-green. These tightly foliaged plants – such as the Lonicera hedges – are used close to the ground, in keeping with a traditional topiary-style approach. Alternatively, some of the more vertical plantings – like the centrally located weeping mulberry, Indian Bean Tree, Juniper Skyrockets, and dwarf lavender – were selected with the nearby bushland in mind. Their less dense, slightly feathered foliage balances the need for formality with a softer, less manicured ambience. At night, the vertical plantings, along with the two-tiered concrete water features, are uplit, casting a magical glow.

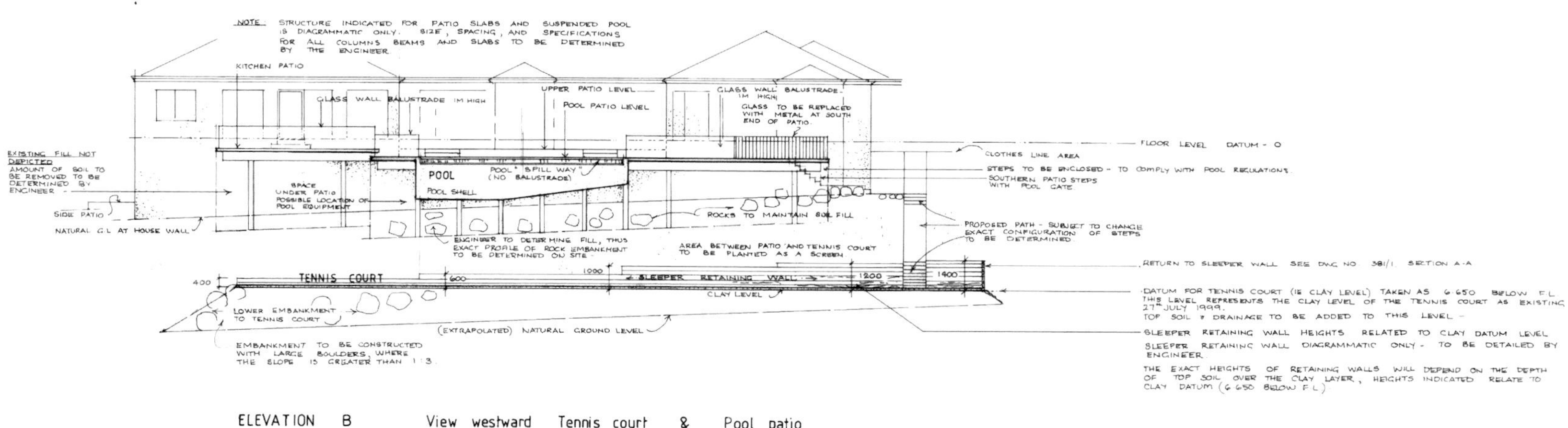

ELEVATION B View westward Tennis court & Pool patio

The two fountains play an important role in connecting the garden to the house. Charcoal grey in colour, they complement the tiled roof. This fosters a sense of belonging without running the risk of the fountains being 'lost' against the exterior facade of the house, which would have happened if a lighter colour had been used. In addition, the subtle sound of cascading water serves as a link to the pool at the rear, which is visible upon opening the front door.

The outdoor environment has been designed specifically to juggle the contradiction between the classical formal elegance of the home and the 'natural', Australian regional setting of the property. This has been successfully achieved by thematically relating the areas adjacent to the home to the house style itself, whilst blending more formal style with the surrounding bushland as one moves further away from the house. The result is that the contradiction has been avoided and the 'setting' of the home is 'appropriate'.

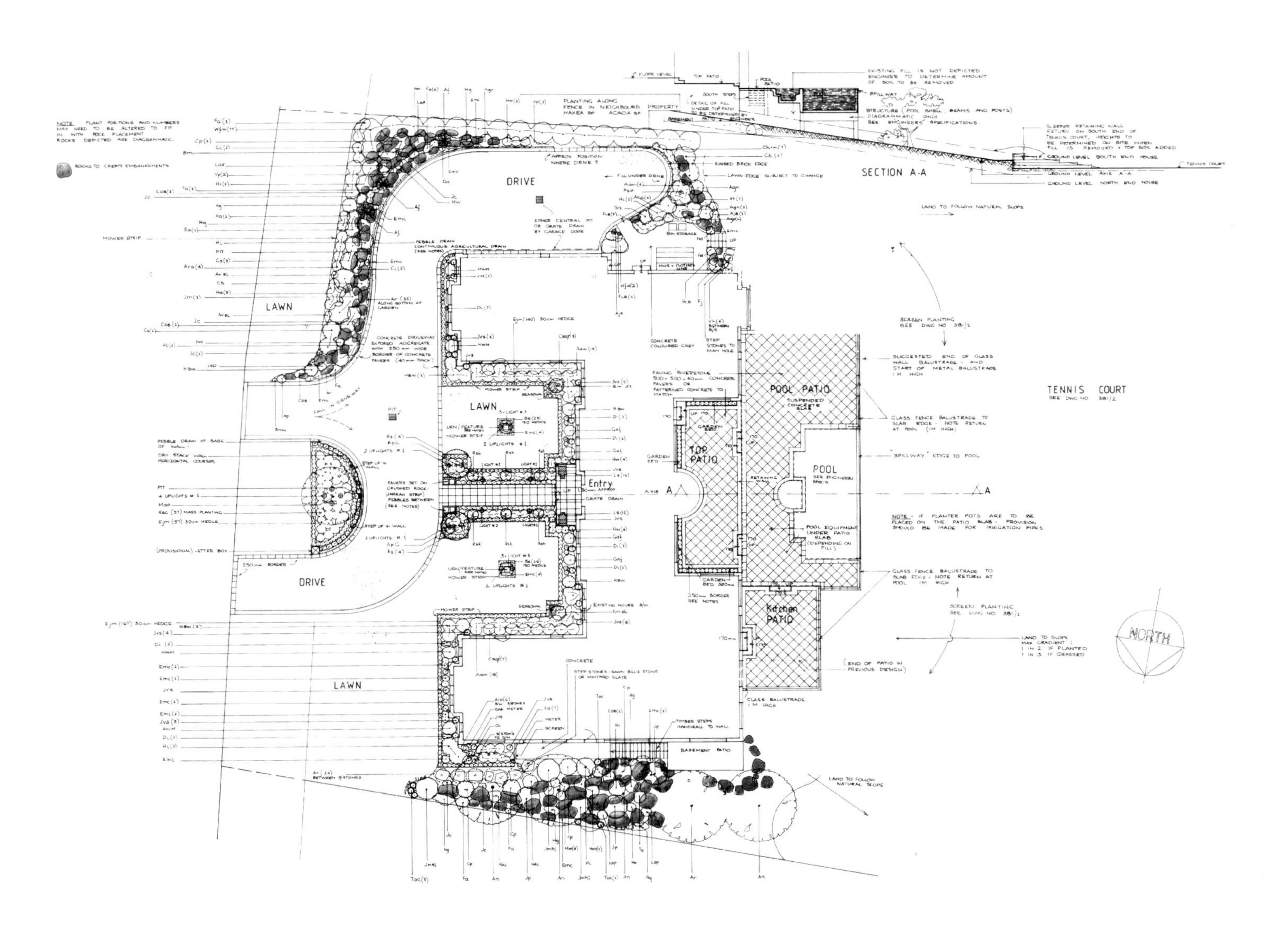

DRIVE
LAWN
DRIVE
LAWN
Entry
TOP PATIO
POOL PATIO
POOL
Kitchen PATIO
TENNIS COURT
SECTION A-A
NORTH

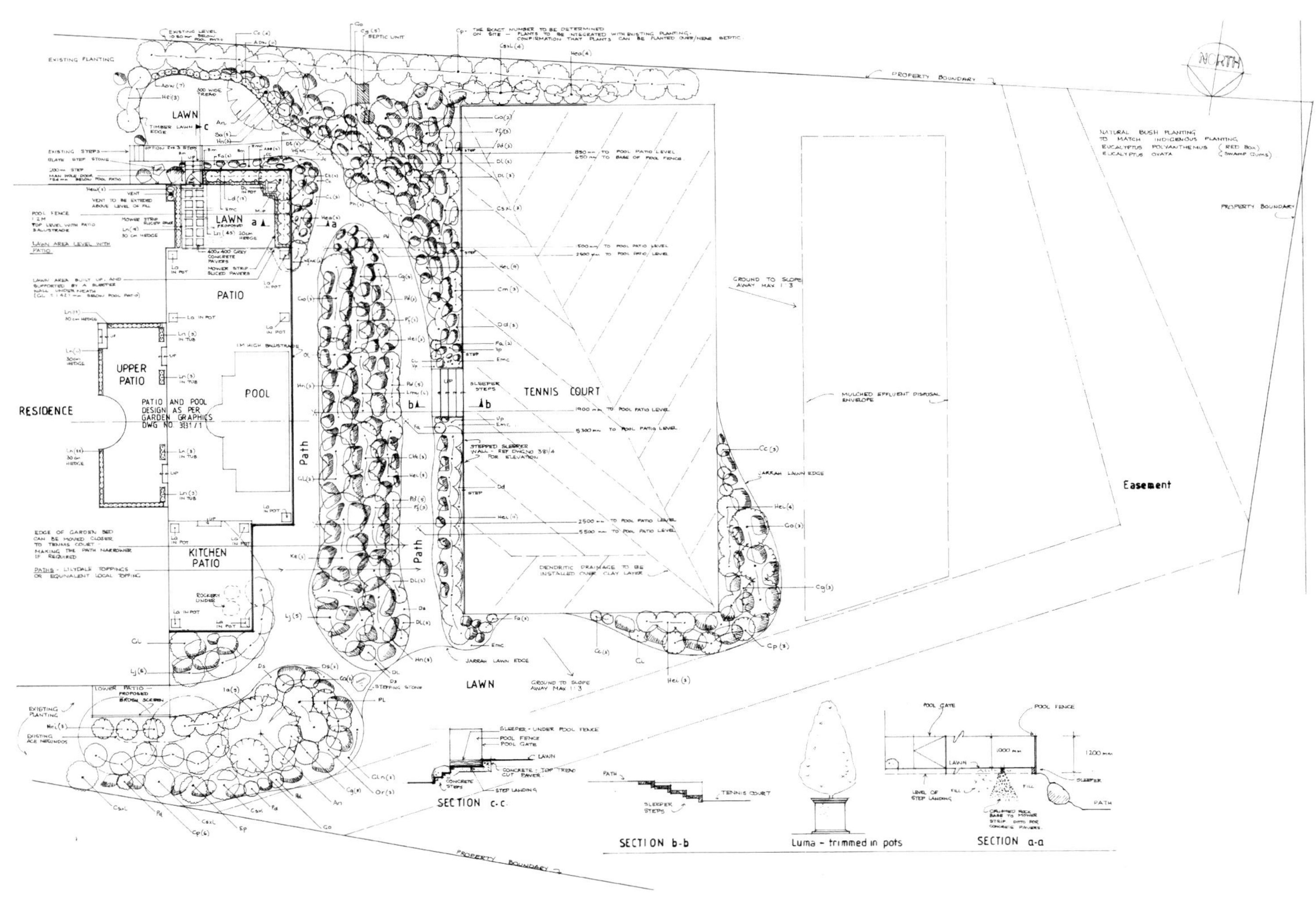

LAWN
LAWN
PATIO
UPPER PATIO
RESIDENCE
POOL
PATIO AND POOL DESIGN AS PER GARDEN GRAPHICS
KITCHEN PATIO
Path
Path
TENNIS COURT
LAWN
Easement
NORTH
PROPERTY BOUNDARY
SECTION c-c
SECTION b-b
Luma - trimmed in pots
SECTION a-a

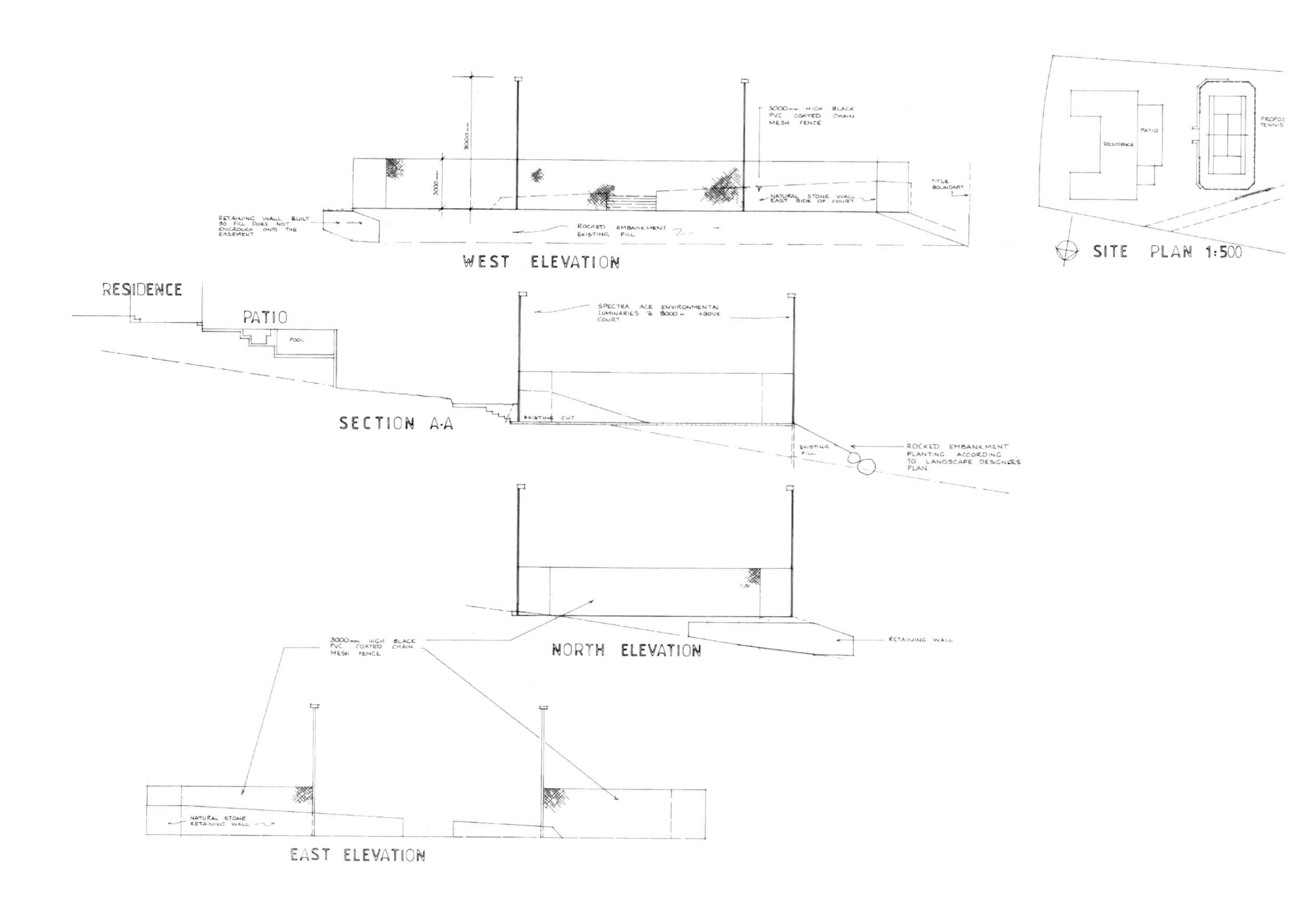
3000mm HIGH BLACK PVC COATED CHAIN MESH FENCE
TITLE BOUNDARY
NATURAL STONE WALL EAST SIDE OF COURT
RETAINING WALL BUILT SO FILL DOES NOT ENCROUCH ONTO THE EASEMENT
ROCKED EMBANKMENT EXISTING FILL
WEST ELEVATION
RESIDENCE
PATIO
SITE PLAN 1:500
RESIDENCE
PATIO
POOL
SPECTRA ACE ENVIRONMENTAL LUMINARIES @ 8000mm ABOVE COURT
SECTION A-A
EXISTING CUT
EXISTING FILL
ROCKED EMBANKMENT PLANTING ACCORDING TO LANDSCAPE DESIGNERS PLAN
3000mm HIGH BLACK PVC COATED CHAIN MESH FENCE
NORTH ELEVATION
RETAINING WALL
NATURAL STONE RETAINING WALL
EAST ELEVATION

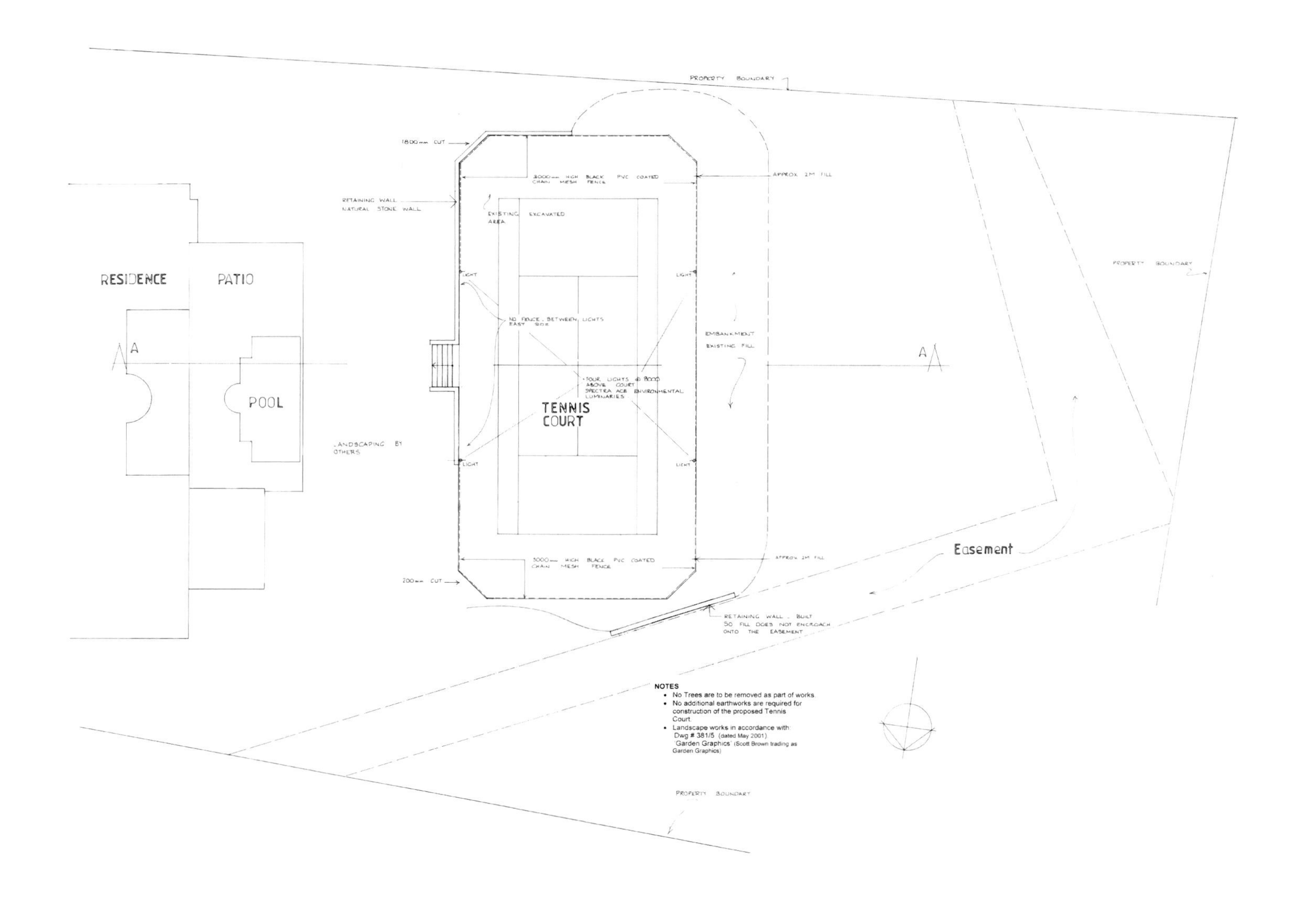
PROPERTY BOUNDARY
1800mm CUT
3000mm HIGH BLACK PVC COATED CHAIN MESH FENCE
APPROX 2M FILL
RETAINING WALL NATURAL STONE WALL
EXISTING EXCAVATED AREA
RESIDENCE
PATIO
PROPERTY BOUNDARY
LIGHT
NO FENCE BETWEEN LIGHTS EAST SIDE
EMBANKMENT EXISTING FILL
A
A
FOUR LIGHTS @ 8000 ABOVE COURT SPECTRA ACE ENVIRONMENTAL LUMINARIES
POOL
TENNIS COURT
LANDSCAPING BY OTHERS
Easement
3000mm HIGH BLACK PVC COATED CHAIN MESH FENCE
200mm CUT
RETAINING WALL BUILT SO FILL DOES NOT ENCROACH ONTO THE EASEMENT
NOTES
No Trees are to be removed as part of works.
No additional earthworks are required for construction of the proposed Tennis Court.
Landscape works in accordance with Dwg # 381/5 (dated May 2001) 'Garden Graphics' (Scott Brown trading as Garden Graphics)
PROPERTY BOUNDARY

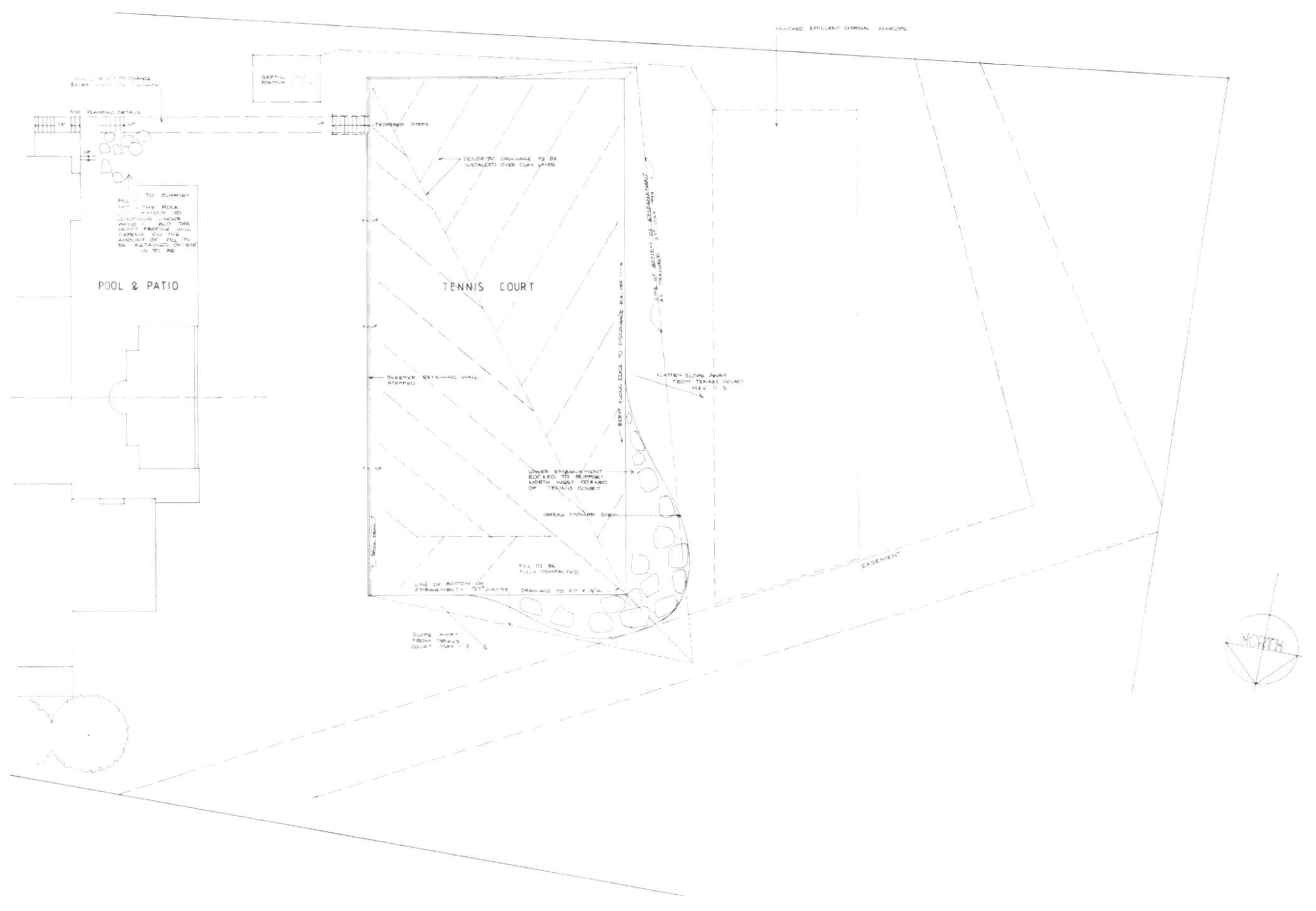
POOL & PATIO
TENNIS COURT
NORTH

Family Garden in Maarjamae

Landscape Architect
Merilen Mentaal

Firm
MentaalLandscapes

Location
Tallinn, Estonia

Area
1,745 m²

Materials
Pinewood decking, treated with oil, concrete stepping stones, concrete paving stones

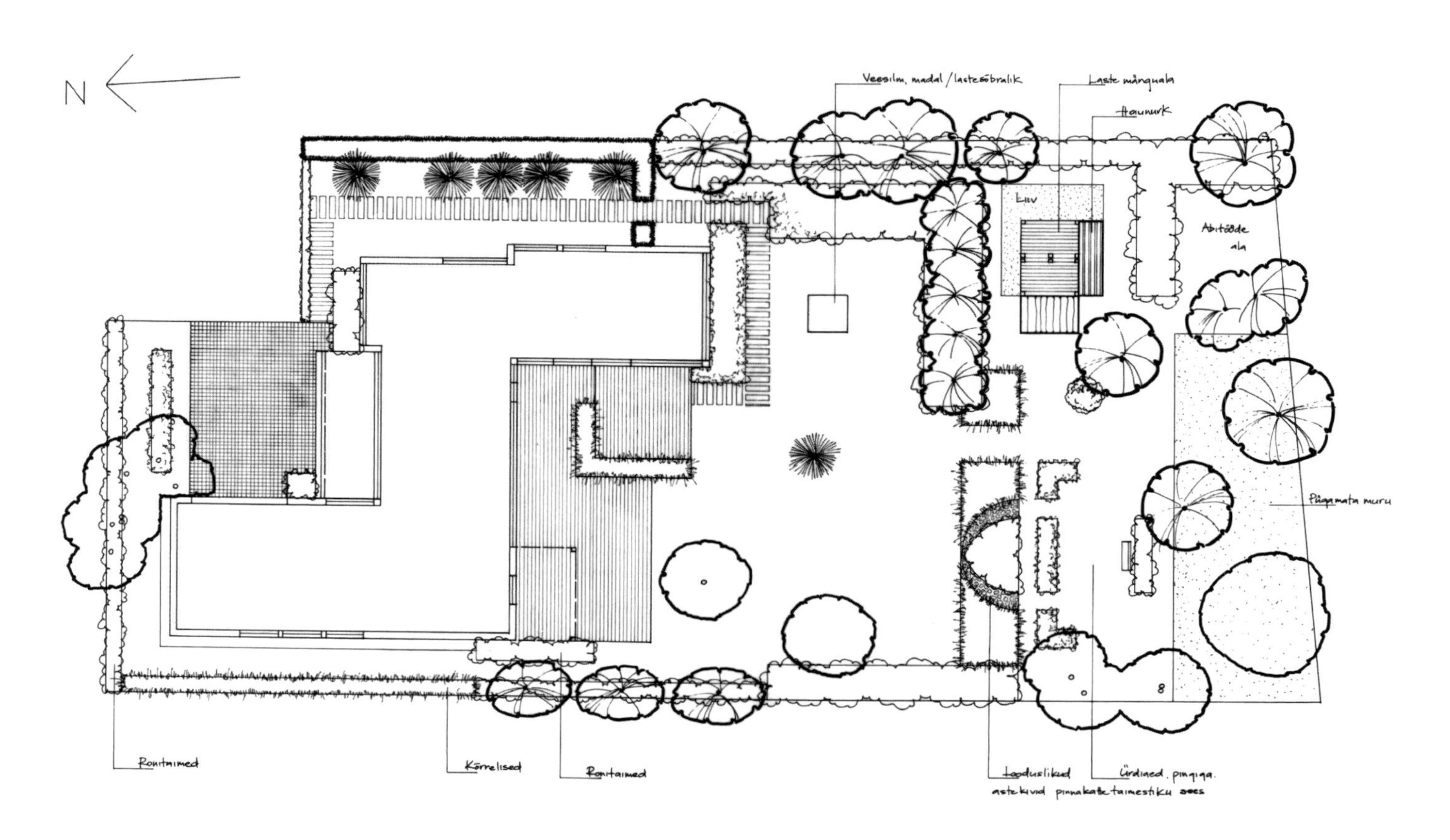

In a quiet residential setting, backed by a hilly forest area, this modern newly-built family house with garden size of 1,745m², asked for a combination of contemporary feeling mixed with lush planting, also considering the needs of playfulness of 3 quite young children.

The design started from the house, creating two rather large decked terraces with one step level change, separated by an L-shaped area filled with Miscanthus grasses and white, pink-tinged tulips in spring. The terraces were designed bold and large to accommodate lounge chairs for sunbathing and after-sauna relaxing, dining table for six, containers with tender plants, leaving also space for children to play. Miscanthus grasses allow privacy and create a lovely, slightly see-through curtain to the garden when seen from indoors.

Tailor-made narrow and long light grey concrete stepping stones connect the front and back gardens, leading to the terrace area. Narrow, dark strip of windows of the indoor swimming-pool area with white wall above it, create rather dramatic background for herbaceous plants to be displayed beautifully to show its shapes and colours throughout the season.

The garden was made especially keeping children in mind. They have their own large play-house with sand area, screened from the house by 5 Malus 'Rudolph' trees. There are several opportunities for climbing, swinging and other play. Back-part of the play-house is a shed/storage-space for seasonal garden tools and other necessities. A slide in between two hazel bushes hides logs for the fire-place, being both, decorative and practical.

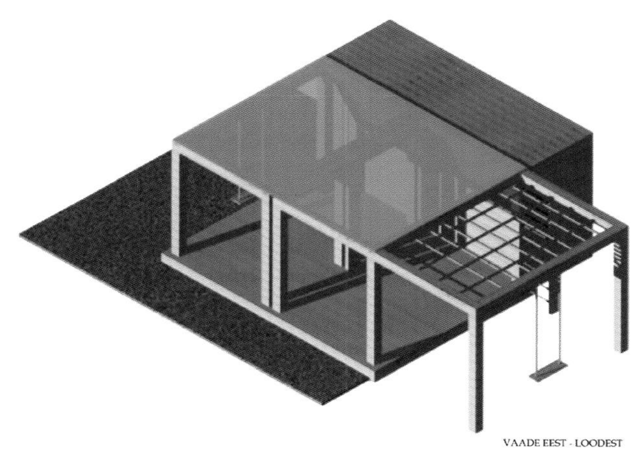
VAADE EEST - LOODEST

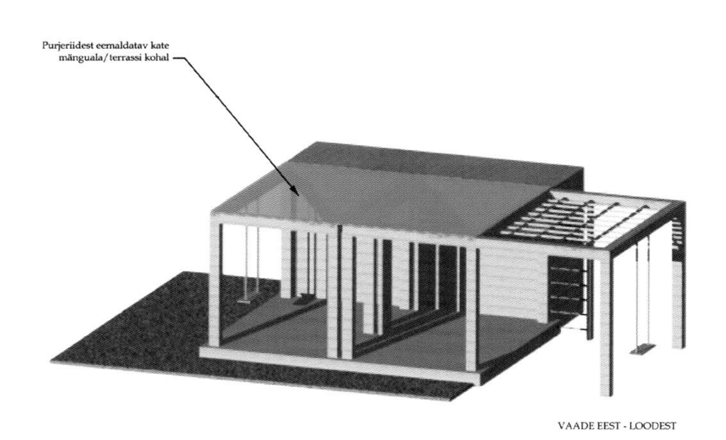
Purjeriidest eemaldatav kate
mänguala/terrassi kohal
VAADE EEST - LOODEST

VAADE TAGANT - LÄÄNEST

VAADE TAGANT - IDAST

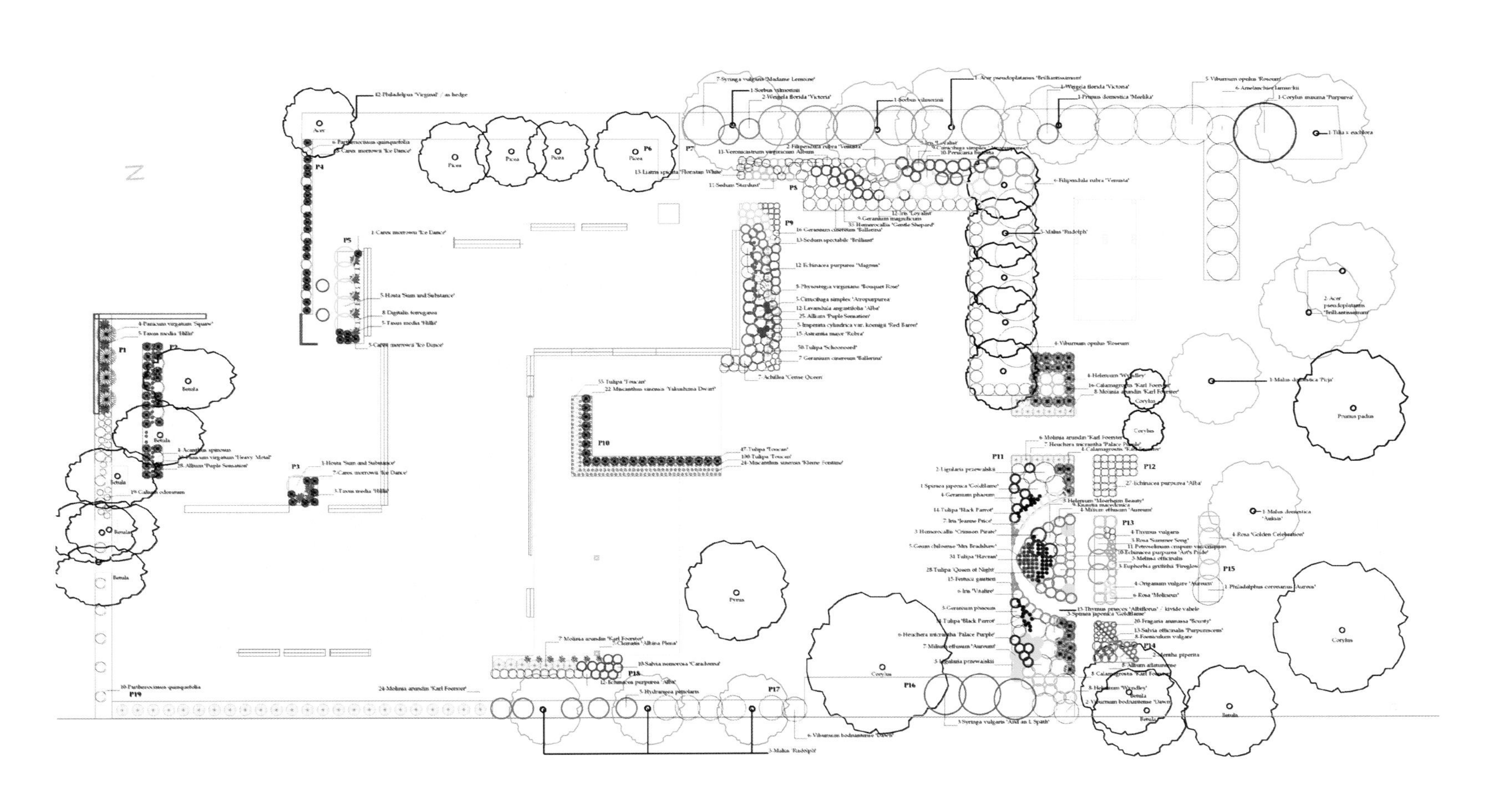

Follers Manor Gardens

Landscape Architect
Ian Kitson Landscape

Firm
Ian Kitson Landscape Architecture and Garden Design

Location
East Sussex, UK

Area
1,800 m²

The garden was one of the channel 4 series 'Landscape Man' projects which presented the gardens from a client perspective. They were also featured on *BBC Gardeners World* in May 2011 (to launch the SGD/RHS gardens open day visits) and this interview and garden footage still provides an excellent design summary and overview.

Key design features of the garden include: the sunken garden, which explores the experience of enclosure and intimacy within larger landscapes as well as providing practical shelter for eating and socializing close to the house; the remodeling of the slope in front of the south elevation by cut and fill to realize the sunken garden; herbaceous displays especially on the south slopes; the wildlife pond, which replaced the tennis court, together with its deck walk which mimics the vegetation forms on the south downs hills; brush strokes of hawthorn hedging acting a structural anchors throughout the garden and again representing elements of the larger landscape within the garden; a clear journey through the garden which both hides and reveals the core garden components. The personality of the garden intentionally ignores the obvious geometric aesthetic of the architecture although there is a sympathetic relationship in the detail use of traditional materials.

The garden has transformed the clients' living experience. They now take every opportunity to be in the garden and consider it an essential component of their home rather than a lifestyle accessory or optional addition.

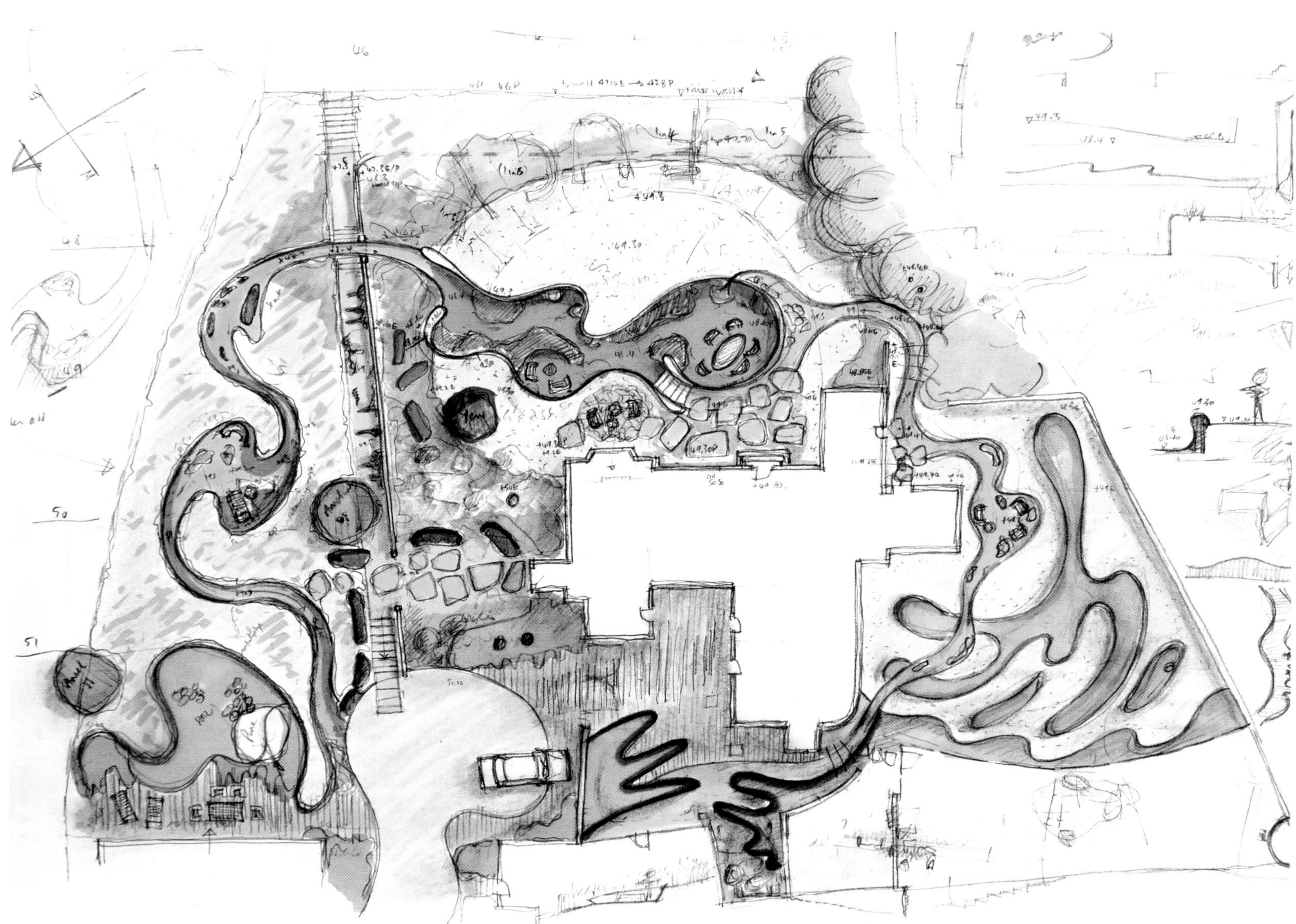

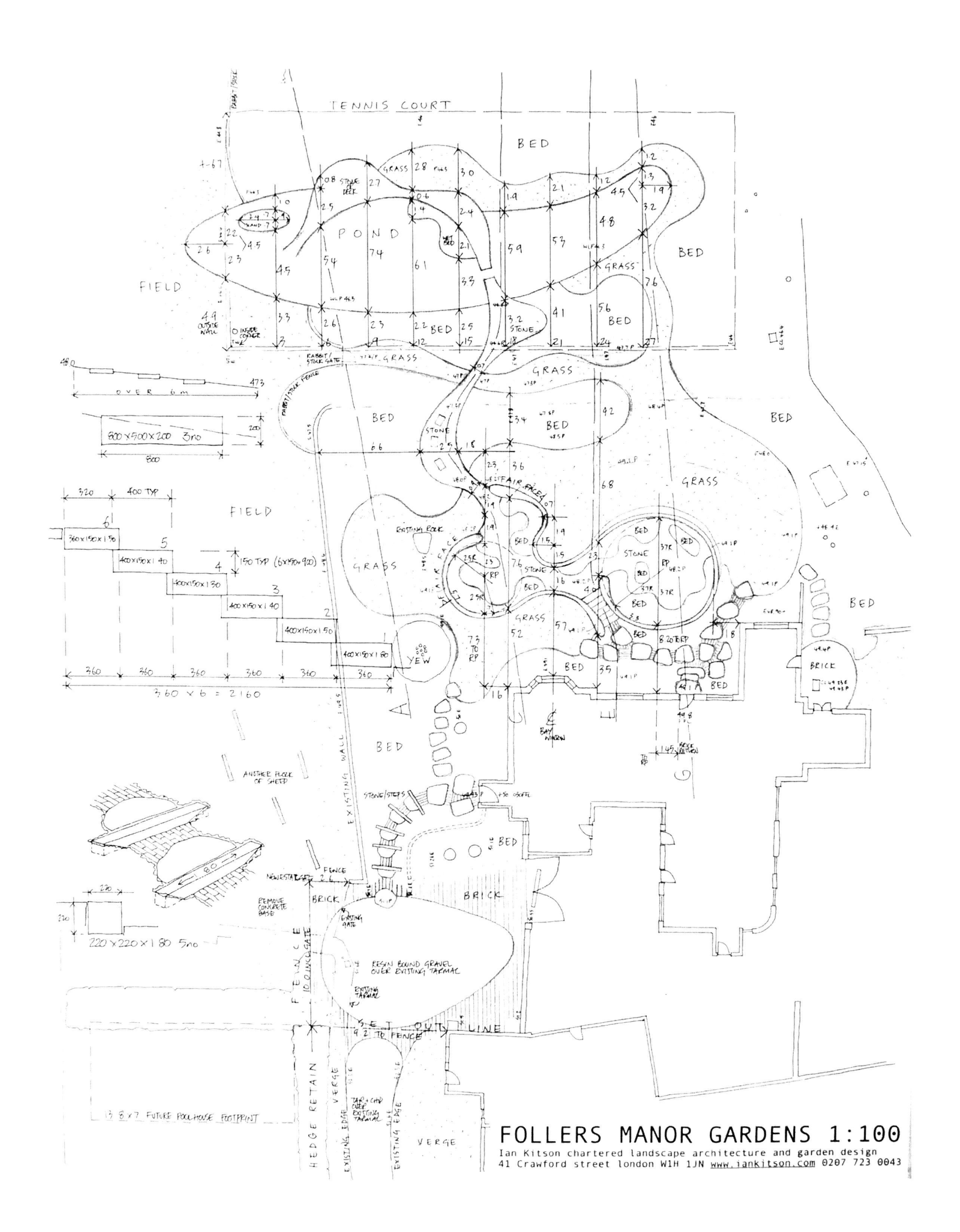
TENNIS COURT
BED
POND
GRASS
FIELD
STONE
YEW
BRICK
EXISTING WALL
OVER 6m
800X500X200 3no
360X150X150
400X150X140
400X150X130
400X150X180
360 X 6 = 2160
220X220X180 5no
RESIN BOUND GRAVEL OVER EXISTING TARMAC
SET OUT LINE
HEDGE RETAIN
VERGE
FOLLERS MANOR GARDENS 1:100
Ian Kitson chartered landscape architecture and garden design
41 Crawford street london W1H 1JN www.iankitson.com 0207 723 0043

Garden for Entertaining in Godalming

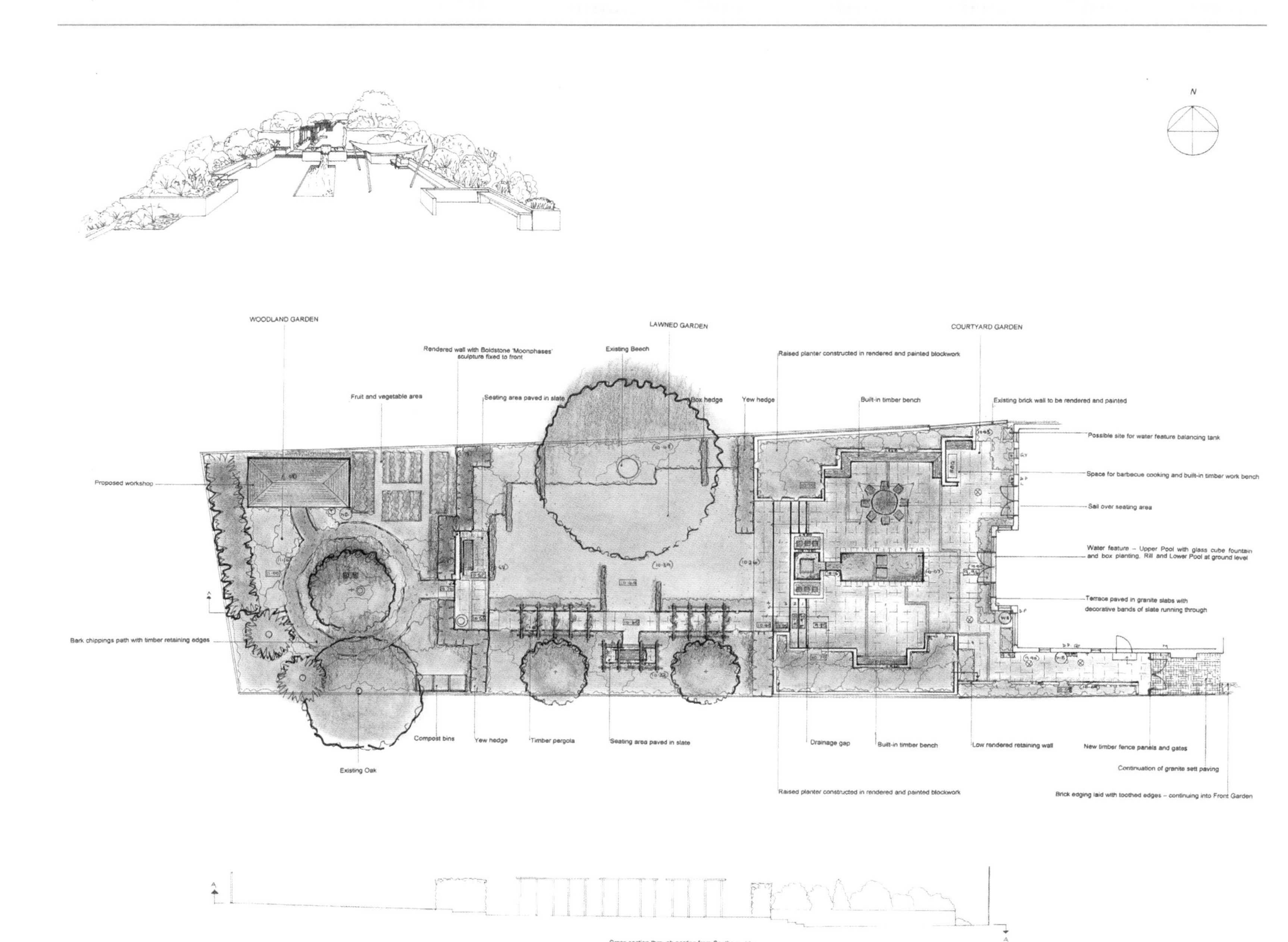

Landscape Architect
Cherry Mills

Firm
Cherry Mills Garden Design

Location
Godalming, UK

The clients wanted a contemporary style for their garden and asked the designer to create a large water feature which could be viewed from its central dining hall. Other requirements were to allocate space for parties near the house, a quiet area for contemplation and a vegetable growing area. Additionally, an important element of the brief was to screen off a large new building which had been constructed on neighbouring land at the end of the garden. The ground slopes upwards away from the house. The designer divides the space into three separate areas and levels – the lower courtyard for entertaining, the lawn area for relaxing and the woodland/vegetable/utility area at the bottom. Each area is separated by tall Yew hedging, which emphasises the change in style but also improves privacy. The paving is grey granite and grey slate. Raised beds have been constructed around the main terrace to improve privacy and give a sense of enclosure. The designer designed and commissioned the glass cube fountain which flows from an upper pool into a shallow pool below, lined with white cobble stones. A double pergola, covered in Roses, Clematis and Jasmine, leads past the lawn to a feature wall, displaying a series of moon phase wall sculptures. The more natural area of the garden is planted up with new trees, including a row of semi-mature Leylandii along the boundary for screening, and woodland plants. A sunny part of the garden is dedicated to vegetable growing.

Kirribilli Residence

Landscape Architect
Terragram Pty Ltd

Architect
Luigi Rosselli Pty Ltd

Location
Sydney, Australia

Photographer
Vladimir Sitta

Architect Luigi Rosselli's approach was to articulate architectural volumes with carving out spaces for small enclosed courtyards and gardens. This approach introduces more light inside the house, but also makes the garden part of the house scenography, unlike the usual approach of placing a large volume in the centre and leaving the garden a completely peripheral experience.

The waterfront garden, with an edgeless definition of lawn creates the illusion of lawn hovering on water. While this garden relies on nautical scenery for its animation, two small spaces rely to a great degree on the detailed resolution and good quality materials and construction. The central courtyard is a simple exercise in geometry that echoes building forms. The sculptural sandstone bench with an ever-changing cross profile is still dominant. This spatial hierarchy will change as the vegetation progresses. Sandstone for the bench has been sourced locally while all other paving is Chinese green granite. A small reflecting pool adds ambience and fills the space with the gentle sound of running water. A small clump of bamboo screens the "uncombed" elevation of the neighbours side, and helps to spatially define courtyard. The paving is designed as permeable, hence there is no visible drainage inlet (part of requirement of authorities for surfaces that reduce stormwater run-off). The fence, normally a utilitarian element, received a bit of quirky treatment – a display of elkhorn ferns, something like trophies used to be displayed in castles and stately homes – with a green twist of course.

The entrance court is a simple space, planted with evergreen plants. In the centre, is a "memory" plant – a mature frangipani that was originally on site, stored elsewhere during building construction, and brought back when the building was completed.

The side passage leading to the water, and consisting of stone interplanted with kidney weed, proves that this Australian native plant and Chinese granite can happily co-exist, and that even such narrow remnants of space can encompass a "garden", visible through the sequence of slot windows from the interior.

Regents Park

Landscape Architect
Kate Gould Gardens

Photographer
Peter Baistow

A small garden links a house and its mews building together. After extensive renovation to the property and the addition of a large glass extension, the garden became far more integral to the internal living space. Accessed from two levels of the house it is visible at all times from the kitchen.

The brief was for an all year round green space with a 'hint of Portofino', cobbled mosaic flooring. Water, Olives, neatly clipped Buxus hedging as well as colourful easy maintenance planting, helps to focus the eye in the garden and not on the adjacent buildings. More traditional materials were used for the paving; new sawn York stone pavers in large format sizes contrast the tiny detail of the mosaic and help to set the garden into its surroundings as a London space.

To create a sense of depth in the garden the mews house wall was clad in mirror polished stainless steel and overlaid with trellis which gives an illusion of space. An unusual glass lantern hangs over the doorway and adds scale at a higher level.

Room with a View

Landscape Architect
Three Sixty Design

Location
Castle Pines, Colorado, USA

Photographer
Michael Peck

The Front Range in Colorado has some of the most beautiful vistas in the country, and maximizing those views not only pleases homeowners, but also increases property values. The landscape architect tackled this particularly tricky project, keeping its spectacular view front and center. Steep grades and exposed bedrock, protected visual corridors and maintaining the feeling of the rugged landscape atop the ridge while weaving in a particularly complex program were the objectives of the project. The result is a design that intertwines clean lines, colour and texture into the landscape beyond.

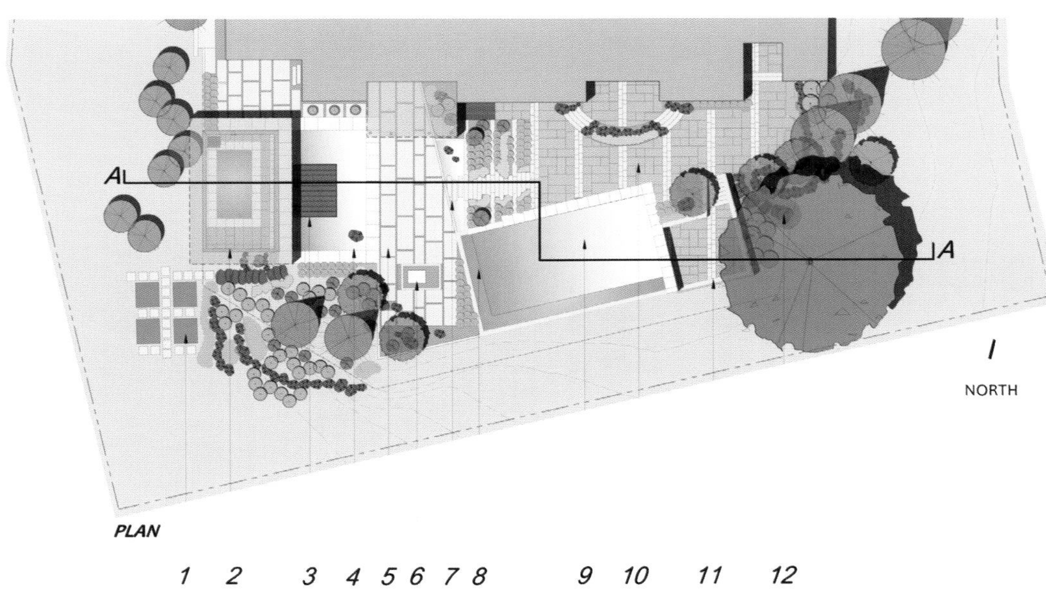

NORTH

ROOM WITH A VIEW

LOCATION: CASTLE PINES, COLORADO
SCOPE: REFLECTING POOL, UPPER AND LOWER TERRACES, VEGETABLE GARDEN, FIRE PIT, WATER FEATURE, DOUBLE SIDED INFINITY POOL WITH WATER WALL, SUNKEN TERRACE, REVEGETATE NATIVE AREAS AND PLANTING

CONSTRUCTION BUDGET: $525,000.00

PLANNING, DESIGN AND CONSTRUCTION TIMELINE: 3 YEARS

LEGEND

1. VEGETABLE GARDEN AND CUT PERENNIAL GARDEN
2. SPA HOUSE
3. IPE DECK
4. REFLECTING POOL
5. LOWER TERRACE (POURED IN PLACE, CONCRETE)
6. FIREPIT (POLISHED CONCRETE)
7. WATER FEATURE
8. WATER WALL
9. INFINITY EDGE POOL
10. UPPER TERRACE (BLUESTONE)
11. SUNKEN TERRACE (BLUESTONE WITH GLASS RAIL)
12. SILOAM STEPS

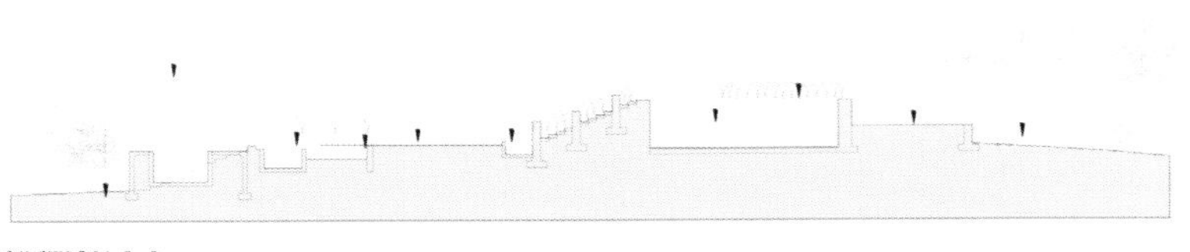

The Contemporary Cotswolds Garden

Landscape Architect
Nick Hendy

Firm
Hendy Curzon Gardens Ltd.

Location
Cotswolds, UK

Plant Designer & Garden Stylist
Adrienne Curzon

Photographer
Adrienne Curzon

Nestled into a Cotswolds village in the UK, this family garden functions on multiple levels that have direct relationships with the rooms of the split-level barn conversion that they connect on to or are viewed from. It was designed to feel like a series of linked spaces.

The Contemporary Cotswolds Garden is multi-faceted but each area and level is unified by the use of a set range of locally sourced materials, planting genre and palette limitation.

A contemporary feel comes from the use of lighting, architectural plant selection, metal structures and the simple shapes of modern furniture, pots and plinths.

Although there are eight levels to the garden it functions across three zones. The first are the courtyard style gardens that wrap around the main living areas of the house. The property entrance, the Kitchen garden and Lounge garden are followed by the Sunken Courtyard which links to the internal den. The second zone is elevated at rooftop level to the house comprising of a very green lawn and topiary garden with a sun-shade area deck and a sun deck. This family area is for entertaining, dining and with a play area and house. The third zone is for work and is at the highest point of the garden. It is reached by a secret path behind an avenue of trees and is not viewable from the house.

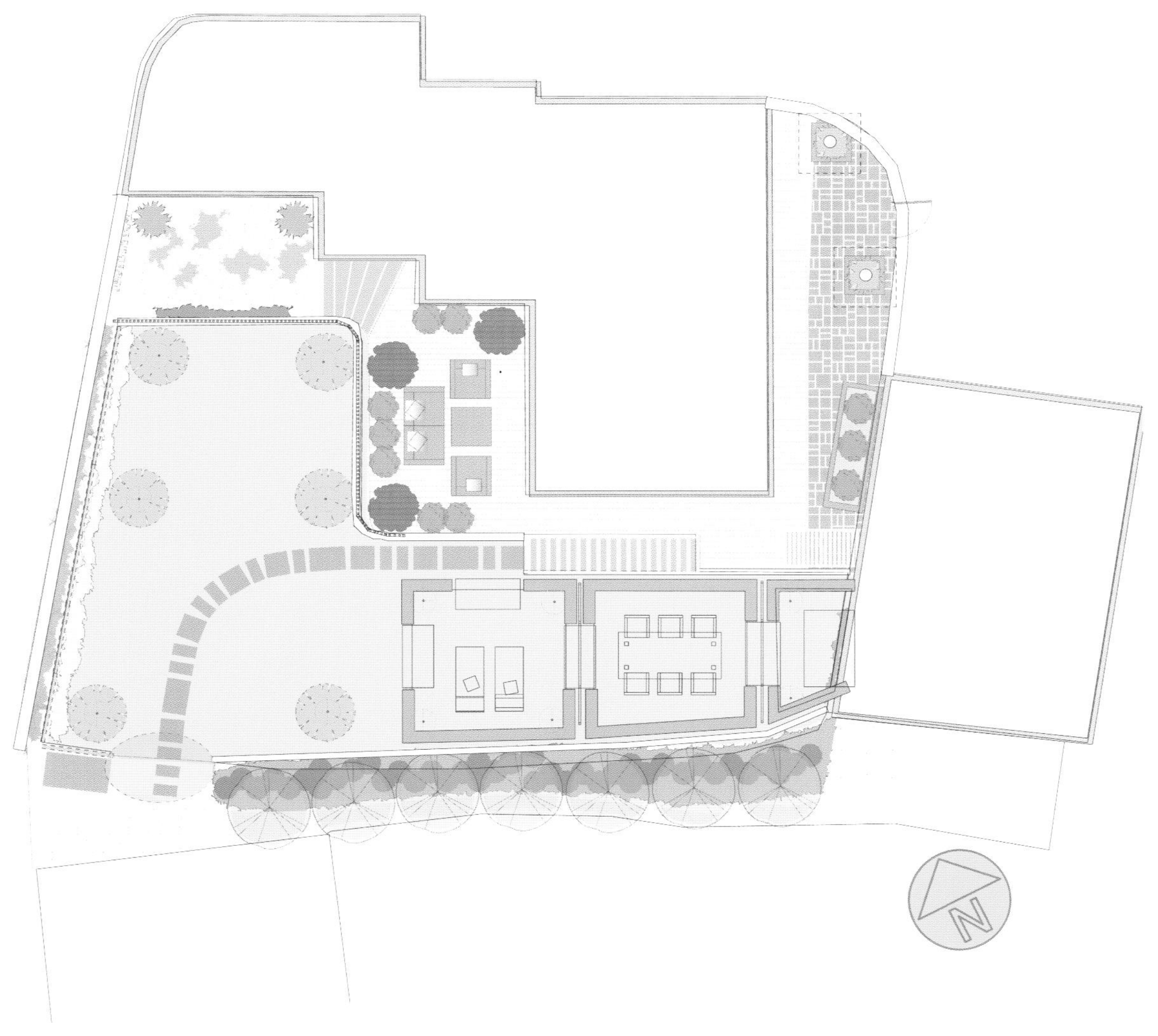
N

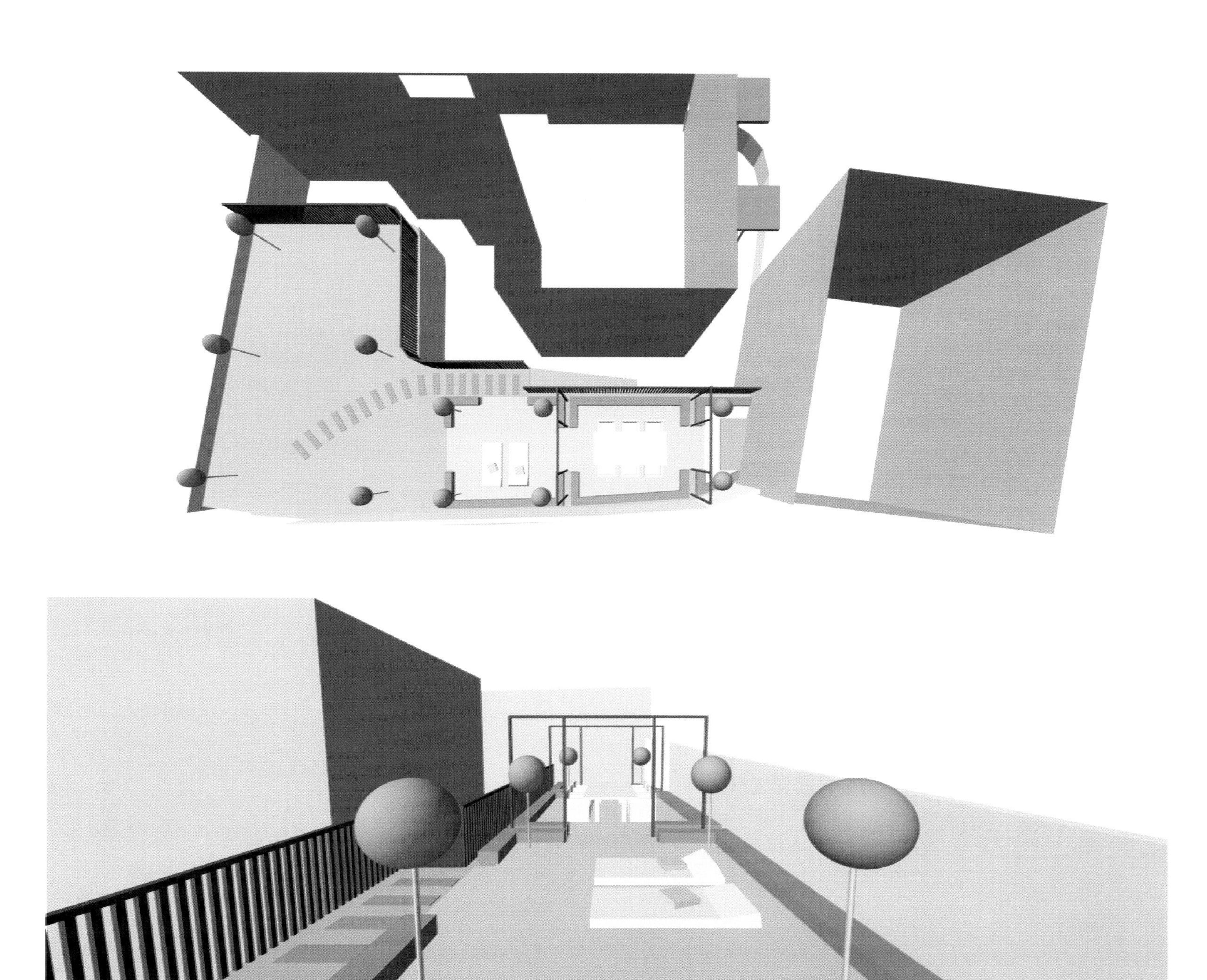

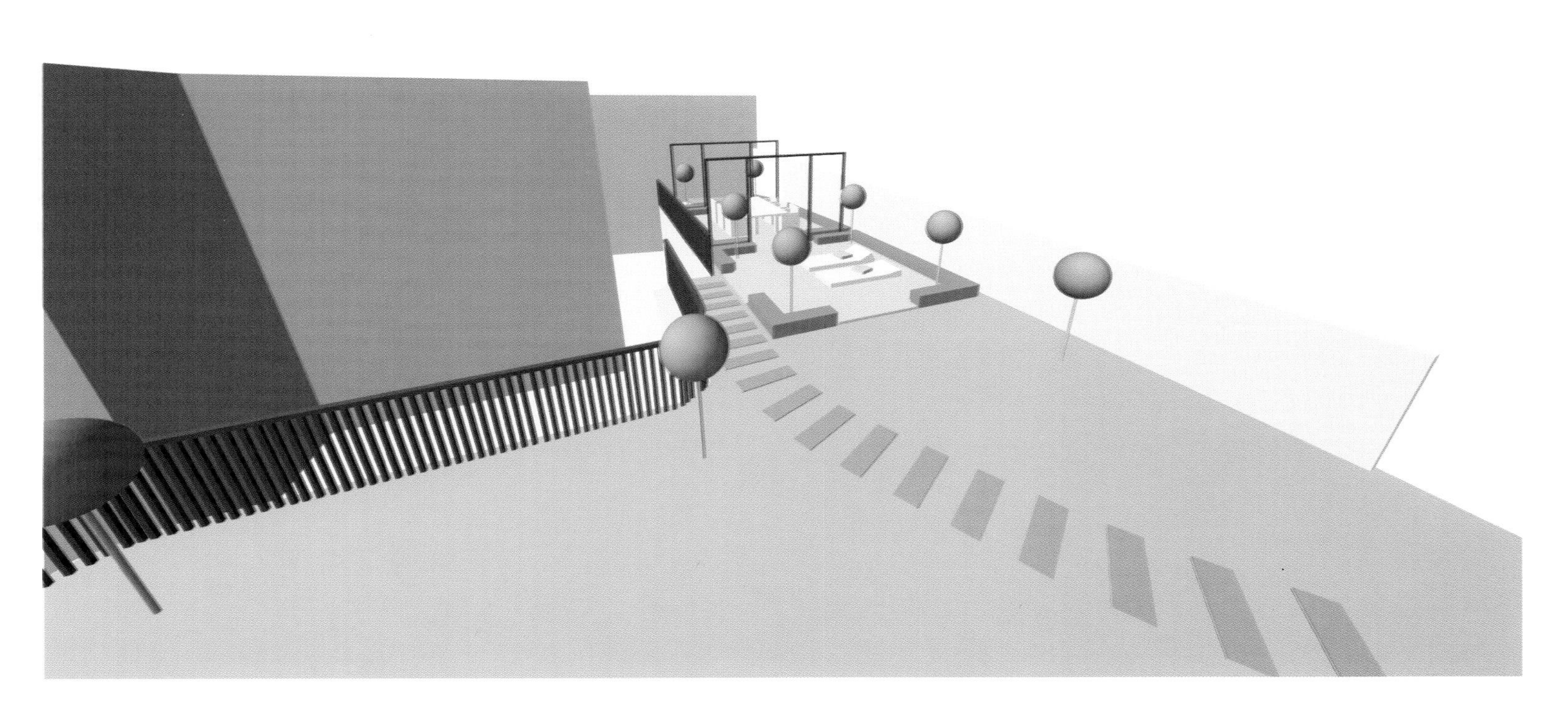

Twadell Residence

Landscape Architect
Rebecca Dye, Hank Helbush,
Kelly Krotcova

Firm
Design Focus Landscape Architecture & Construction

Architect
Min Day

Location
Silicon Valley, USA

Photographer
Lauren Devon, Elizabeth Murray

This contemporary garden was a challenge because of the steep slopes and frequent elevation changes. However, elevation became an asset in the design of the vanishing edge pool. A 1.83m drop at the end of the pool allowed for a dramatic spillway and a remarkable view of the surrounding hills and Silicon Valley. The contemporary pool area was designed to be in stark contrast to the natural environment and be a recreational oasis surrounded by natural views. Large cast concrete pavers provided the surface to accomplish the clean modern look and relate to the architecture of the house. The colour of the pool tiles and plaster were selected to mimic the color of the sky. The designers also painted the arbor a striking orange to play against the sky, and the purple blooming wisteria in a playful way.

Because one of the owners was handicapped, a contemporary ramp was designed by Min Day to provide access to the pool area. They selected artificial turf so the lawns could be wheelchair accessible, and reduce water consumption in the dry environment.

The clients requested unusual plantings such as kniphofia and aloes and wanted plants with interesting shapes as well as plants that featured yellow flowers. Most of the plants included in the design were low water varieties. Another focus of the planting was to create seasonal colour throughout the year. The designers used the planting to frame the views and either contrast or relate to the native surroundings.

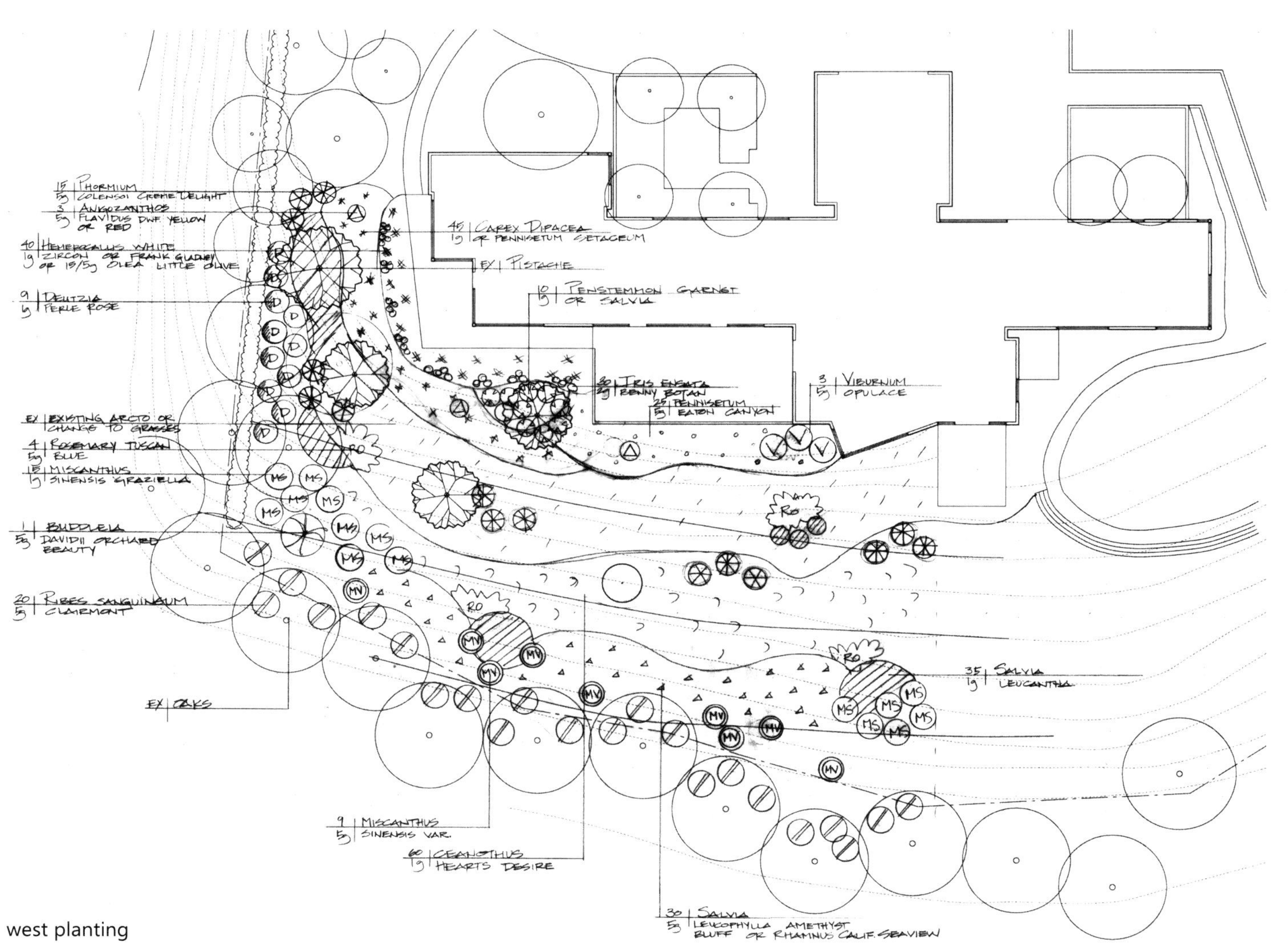

west planting

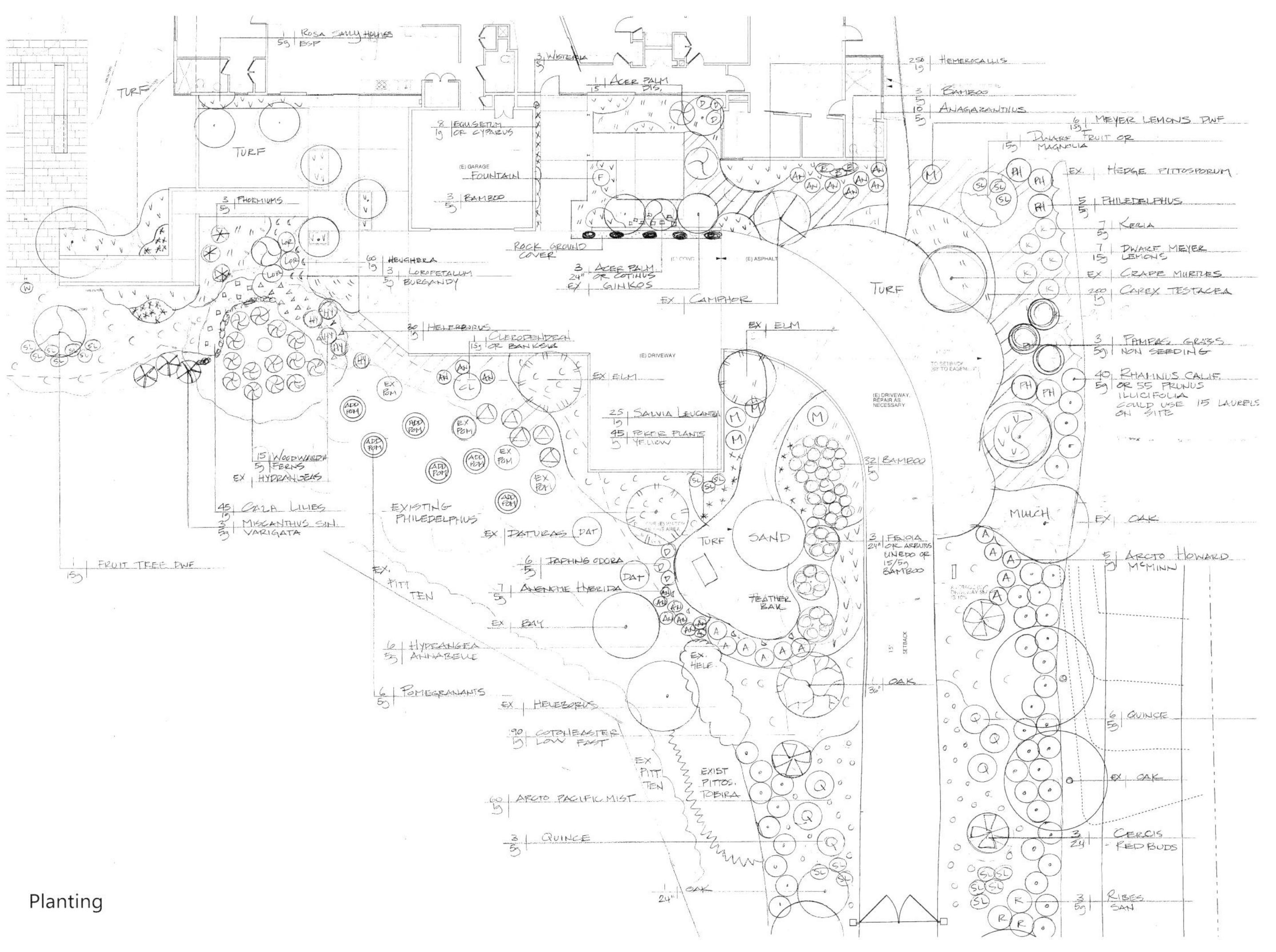

TURF
ROSA SALLY HOLMES
WISTERIA
ACER PALM DIS.
HEMEROCALLIS
BAMBOO
MEYER LEMONS DWF
HEDGE PITTOSPORUM
PHILEDELPHUS
KERIA
DWARF MEYER LEMONS
CRAPE MURTLES
CAREX TESTACEA
PAMPAS GRASS NON SEEDING
EQUISETUM OR CYPRUS
FOUNTAIN
PHORMIUMS
HEUCHERA
LORIPETALUM BURGANDY
ROCK GROUND COVER
ACER PALM OR COTINUS
GINKOS
CAMPHOR
ELM
HELLEBORUS
SALVIA LEUCANTHA
WOODWARDIA FERNS
HYDRANGEAS
CALLA LILIES
MISCANTHUS SIN VARIGATA
EXISTING PHILEDELPHUS
DATURAS
SAND
MULCH
OAK
ARCTO HOWARD McMINN
FRUIT TREE DWF
DAPHNE ODORA
FEATHER BAY
HYDRANGEA ANNABELLE
POMEGRANANTS
HELEBORUS
COTONEASTER LOW FAST
EXIST PITTOS. TOBIRA
ARCTO PACIFIC MIST
QUINCE
CERCIS RED BUDS
RIBES SAN

Planting

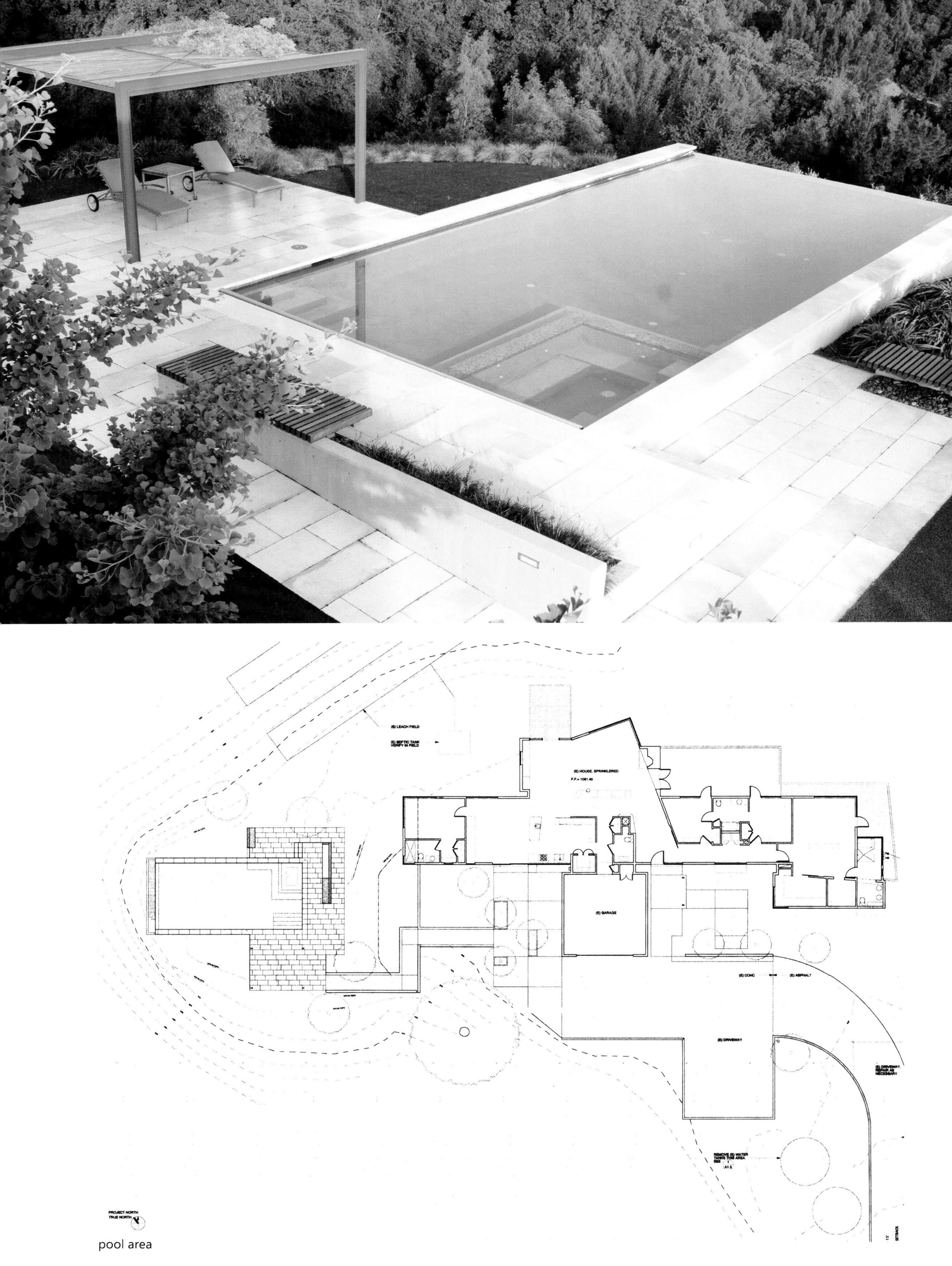

(E) LEACH FIELD
(E) SEPTIC TANK VERIFY IN FIELD
(E) HOUSE, SPRINKLERED
F.F.= 1081.40
(E) GARAGE
(E) CONC
(E) ASPHALT
(E) DRIVEWAY
(E) DRIVEWAY, REPAIR AS NECESSARY
REMOVE (E) WATER TANKS THIS AREA SEE 1 A1.3
15'
SETBACK
PROJECT NORTH
TRUE NORTH

pool area

pool area planting

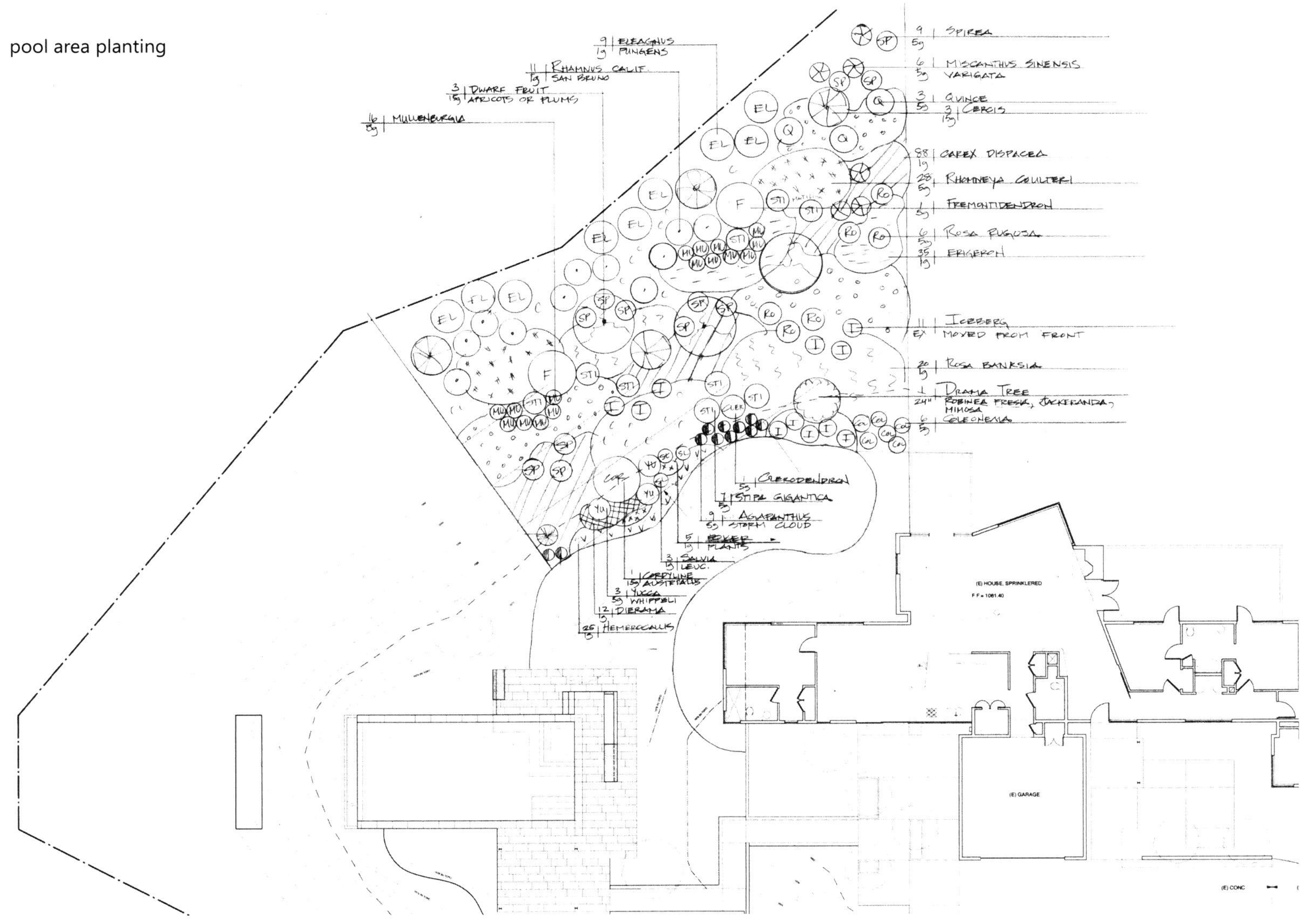

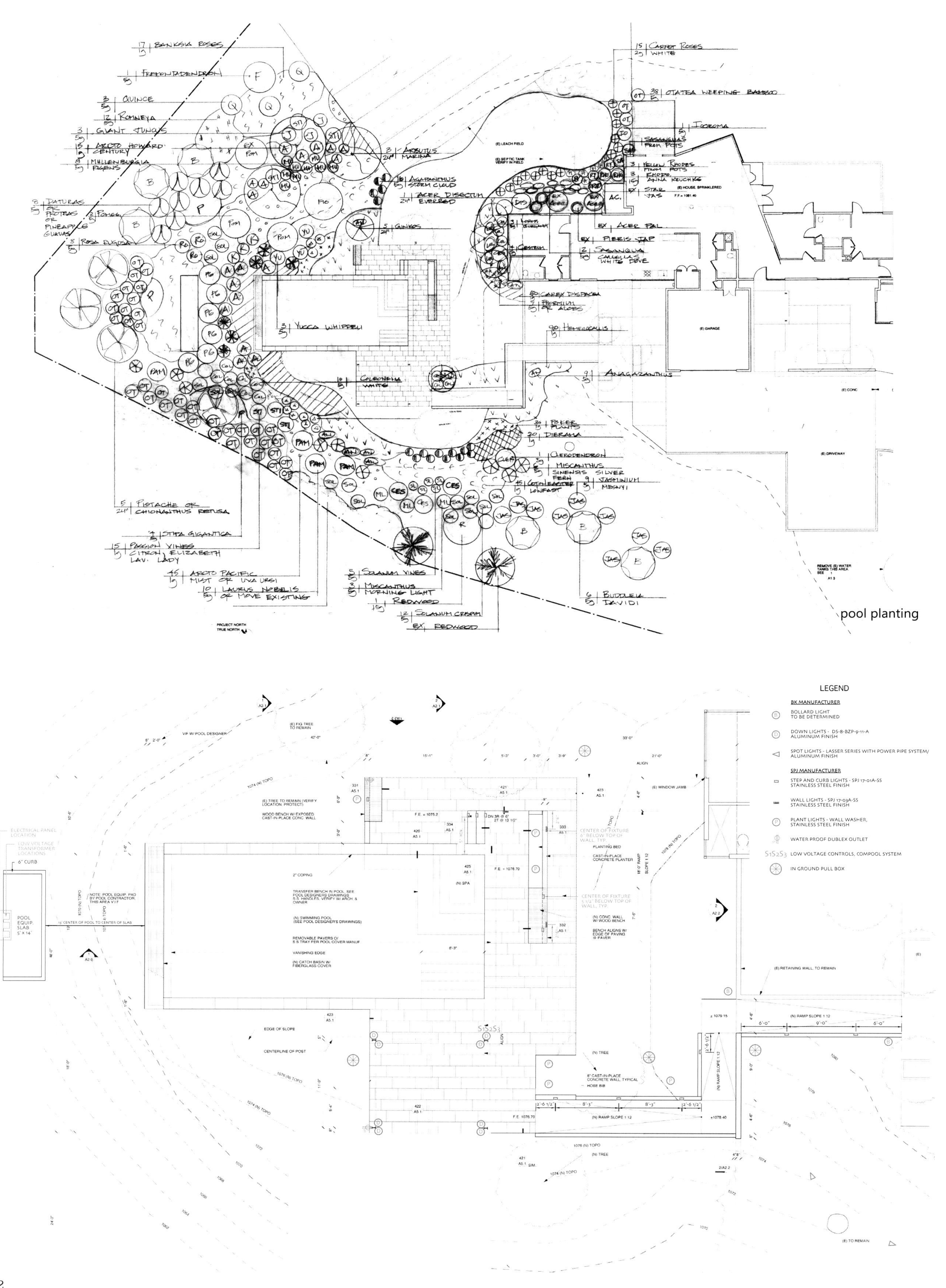

pool planting

Large Family Garden in Guildford

Landscape Architect
Cherry Mills

Firm
Cherry Mills Garden Design

Location
Godalming, UK

The clients, who had recently purchased this Listed house, did not feel the layout of the garden provided them with enough play space for their growing family or reflect their preference for a design with clean lines and a more contemporary feel. The planting adjacent to the house was so tall one could not see the garden from the ground floor rooms. Concrete driveway pavers had been used for paths throughout the garden. An existing L-shaped pond was in need of refurbishment.

This new design has opened up sight lines through the garden, enlarged the lawn and created new paths and neater planting areas. A large slate waterfall is now the focal point at the end of the garden. Beside it is a new play area with swings. The pond has been restored and a new hardwood deck has been constructed around it. There is a new dining terrace with feature sculpture on a wall. An area of planting across the back of the house has been replanted with a Box parterre, designed around existing large Box balls. Some existing walls and entrances were altered to create a pretty courtyard, complete with custom made wrought iron gates. During the build process a putting green with artificial turf was incorporated into the main lawn area. A comprehensive lighting scheme has been installed which brings the property and its new features to life at night.

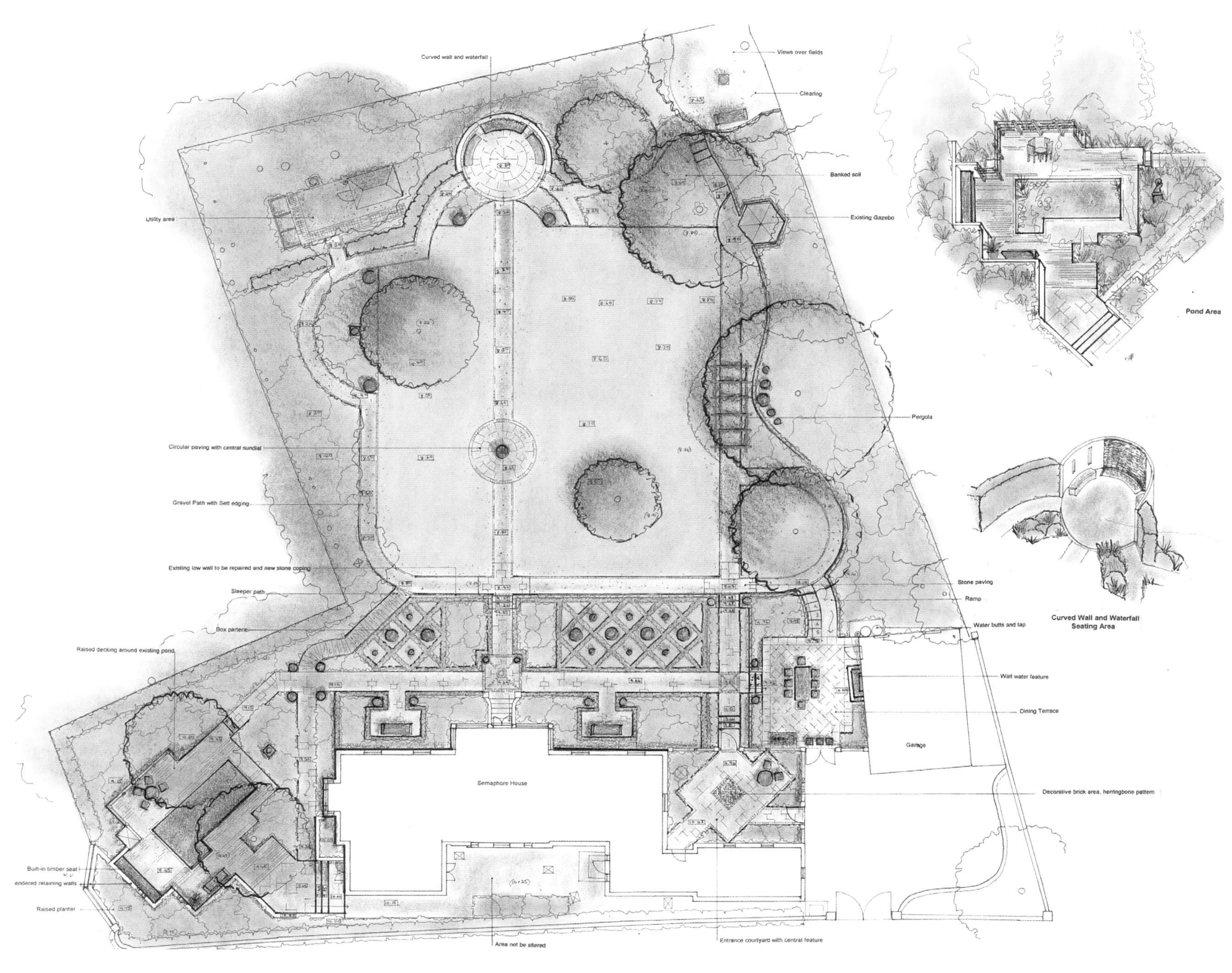

Mostra Black 2011

Landscape Architect
Alex Hanazaki

Firm
Alex Hanazaki Paisagismo

Location
São Paulo, Brazil

Area
340 m² and 158 m² of vertical garden

Photographer
Alex Hanazaki, Marcos Lima, Yuri Seródio

Innovating the concept of vertical garden, a wall of 7m was created with a surprising composition which has joined simple vegetation and cutting edge technological resources like image projection. In the garden, beyond the vegetation were used materials such as volcanic stones Hitam and Hijau imported from Indonesia, which have brought to the pool, a large mirror pool and walkways. Common materials were also used differently, as the iron that coats all gazebo and pergolado ceiling with steel structure.

Rye

Landscape Architect
Scott Brown Landscape Design

Pool
Custom-built, reinforced concrete, waterline ceramic tiles with pebble render interior

Pool Cover
Automatic, retractable cover (Remco)

Pool Coping
Sawn bluestone (China)

Retaining Walls
A/Masonry, clad with block-formed bluestone
B/ Timber sleepers (palisade)

Decking
Timber- Untreated Jarrah (Western Australia)

Building/Pavilion
Timber construction, Timber wall Cladding (Weather resistant)

Pool Fence
Frameless glass and Aviary mesh

Barbeque
Stainless steel, built into a customised timber-clad bench
Tiled splash- back (BBQ)
Hand- made mosaic using ceramic tiles

Outdoor Lighting
Low voltage (Halogen and LED)
Hunza (N.Z.), Megabay (N.Z.), Lumascape (AUST) and Castlight

Location
Mornington Peninsula, Victoria, Australia

Nestled behind the sand-dunes, amongst coastal tea-tree scrub, this weekend 'getaway' on Melbourne's Mornington peninsula is a delightful retreat from the hustle and bustle of the city and the suburbs, and the pressure of deadlines that they represent.

The home is designed in a simple architectural style, with attention being paid to the need for minimal maintenance. The single story home is clad in vertical lined weather-proof cladding, with the roof being hidden by the parapet walls. This allows the home to blend into the landscape, rather than dominate it.

As such, Andrew and Zara wanted the design of the garden, the pool and the alfresco pavilion and entertaining area to follow a similar theme...that of simply fitting in with the existing landscape and vegetation, rather than dominating it.

The Alfresco pavilion is free standing, and has an extensive array of features because it is situated some distance from the home, and as such needs to be largely self-sufficient. The aim has been to create a 'destination' with the pavilion also providing protection from the prevailing south west winds. It is clad in the same exterior cladding as the home, and houses the bar, sink, fridge, cupboards, and seating, toilet and shower.

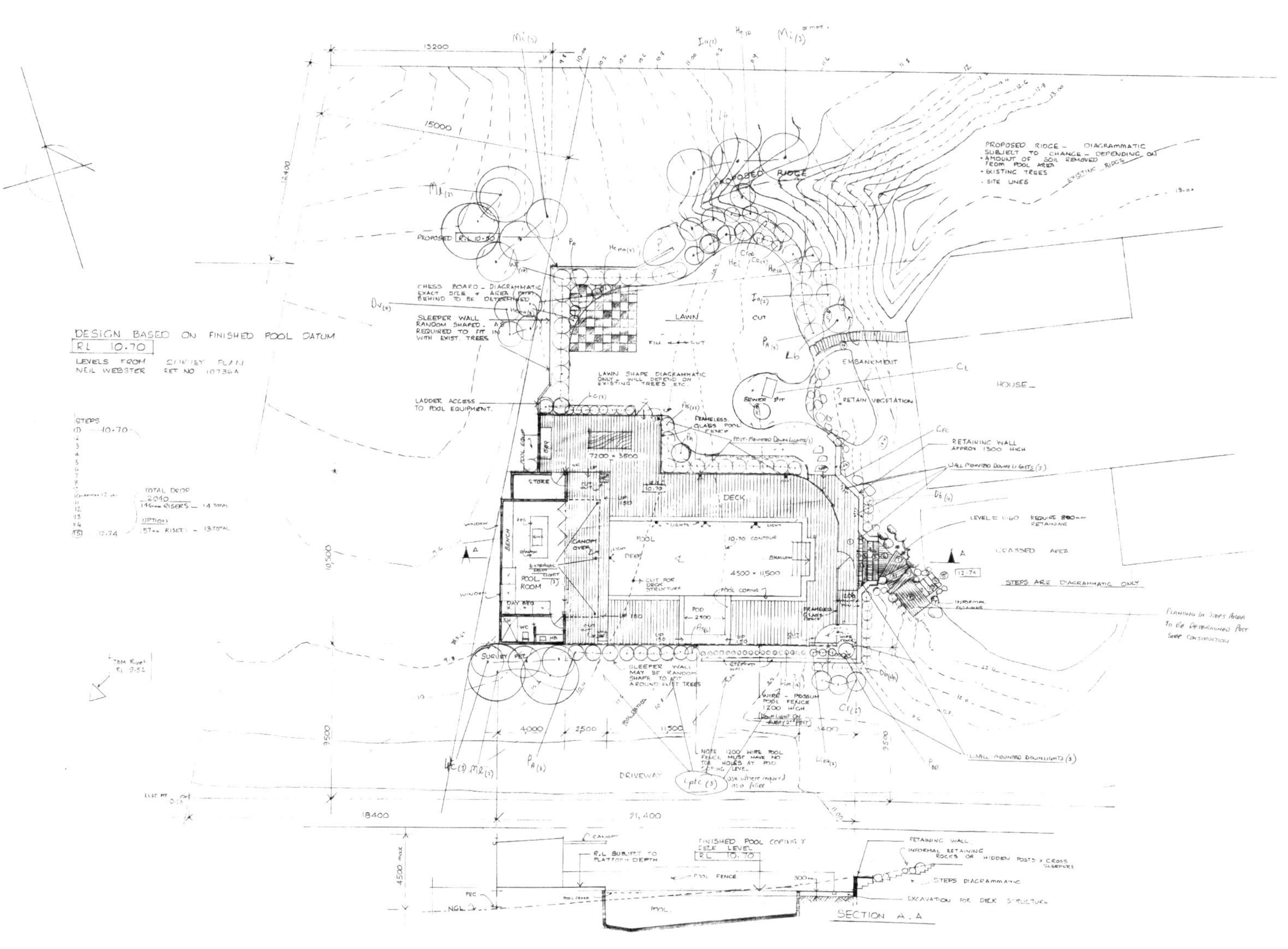
DESIGN BASED ON FINISHED POOL DATUM
RL 10.70
PROPOSED RIDGE
LAWN
HOUSE
EMBANKMENT
RETAIN VEGETATION
SEWER PIT
RETAINING WALL
APPROX 1500 HIGH
DECK
POOL
4500 × 11500
7200 × 3800
POOL ROOM
DAY BED
STORE
STEPS ARE DIAGRAMMATIC ONLY
DRIVEWAY
SECTION A.A

Private Garden nº147 Urbanización Los Lagos Madrid

Landscape Architect
laND30

Firm
laND30

Location
Los Lagos, Madrid, Spain

Area
2,500 m²

Photographer
A-cero

This garden occupying 2,500m²,is located in a private urban development near Madrid, and was planned to serve a modern style villa surrounding it with a pleasurable environment. The garden highlights the architecture and follows the same modern stylistic pattern.

A large area composing a lawn was reserved for human activity and serves as a visual canal to the soothing presence of a water sheet that doubles as a recreational swimming pool.

Sculptural elements are scattered punctuating the space with focal points and suggesting a natural randomness to the environment. These elements vary from Corten steel sculptures to rocks, to notable trees.

The garden is bordered by slopes and vegetation to allow for a more intimate space and there are also smaller isolated spaces that function as scenery for the interior of the house.

The selection of vegetation was very important for this project since it had to be adapted to a climate with harsh winters causing frosts and dry summers lacking water.

The garden can be divided into four main zones: Lawn; The slopes; Interior Courtyard; House entrance.

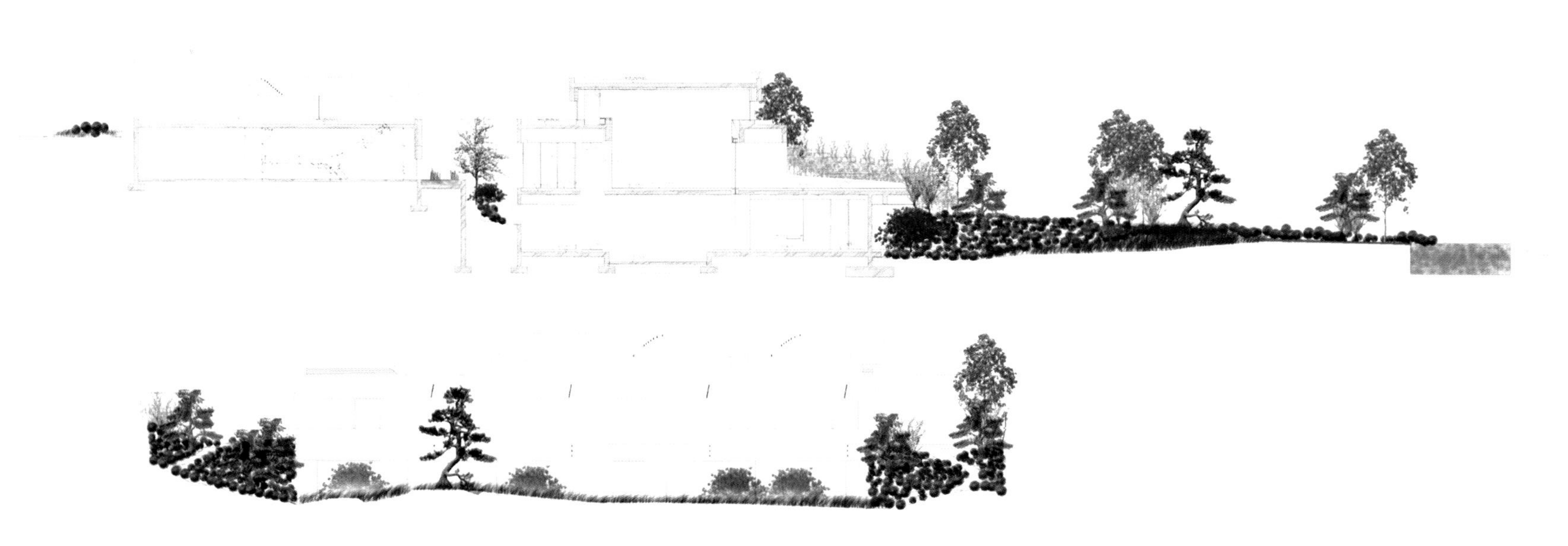

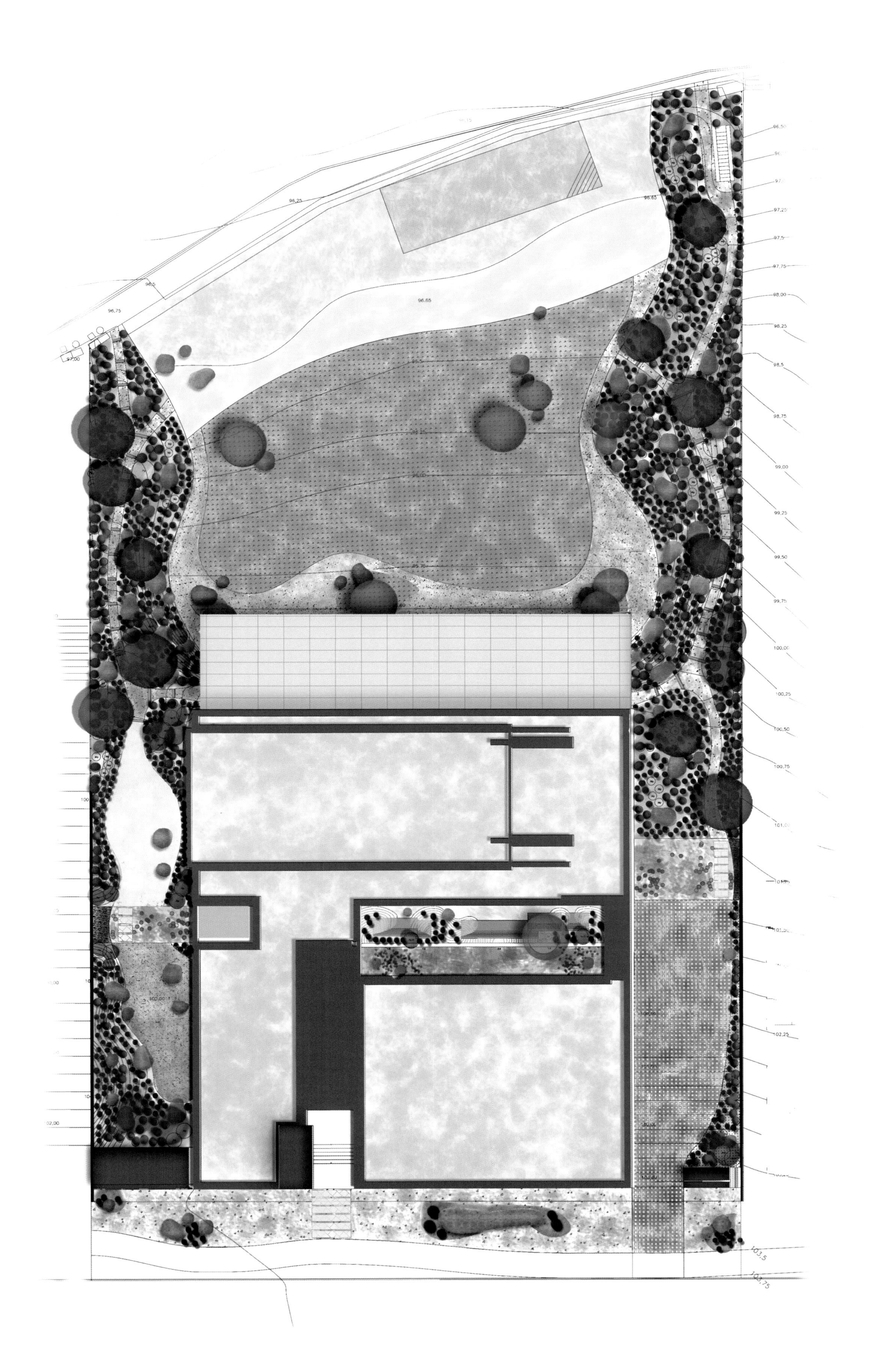
96,25
96,65
96,65
96,75
97,00
97,25
97,5
97,75
98,00
98,25
98,5
98,75
99,00
99,25
99,50
99,75
100,00
100,25
100,50
100,75
102,00
102,25
103,5
103,75

Private Garden nº108 Urbanización Los Lagos Madrid

Landscape Architect
laND30

Firm
laND30

Area
2,500 m²

Architect
A-cero

Location
Madrid, Spain

Photographer
A-cero

The landscape architecture project for this Villa in urbanización Los Lagos, Madrid, was made in collaboration with A-Cero architects and strived to achieve a modern and clean look. Special attention was given to frame and accommodate the architecture while providing an interesting dialog between hard and soft surfaces, geometric and organic shapes as well as built and grown material.

The space is spotted with rocks of different sizes, some of them enormous, arising from a large area of evergreen grasses. This area occupies most of the intervention, giving harmony to the garden and creating spaces that form different paths and seating areas.

A large flat prairie, to the south, surrounded by the house, large shapes of grass and a small lake, serves as the main usage area. Adjacent to the pool there is a chill-out area, next to a circular sandbox simulating a Zen garden, similar to those created in the main entrance area where slate monoliths arise from a combed fine gravel surface.

The tree layer consists of various species forming naturalized forests. Birches were planted in the lake area, while sweetgum separates the meadow from the chill-out zone. While groups of apple trees and strawberry shrubs, recovered from the fields, scatter the entire surface.

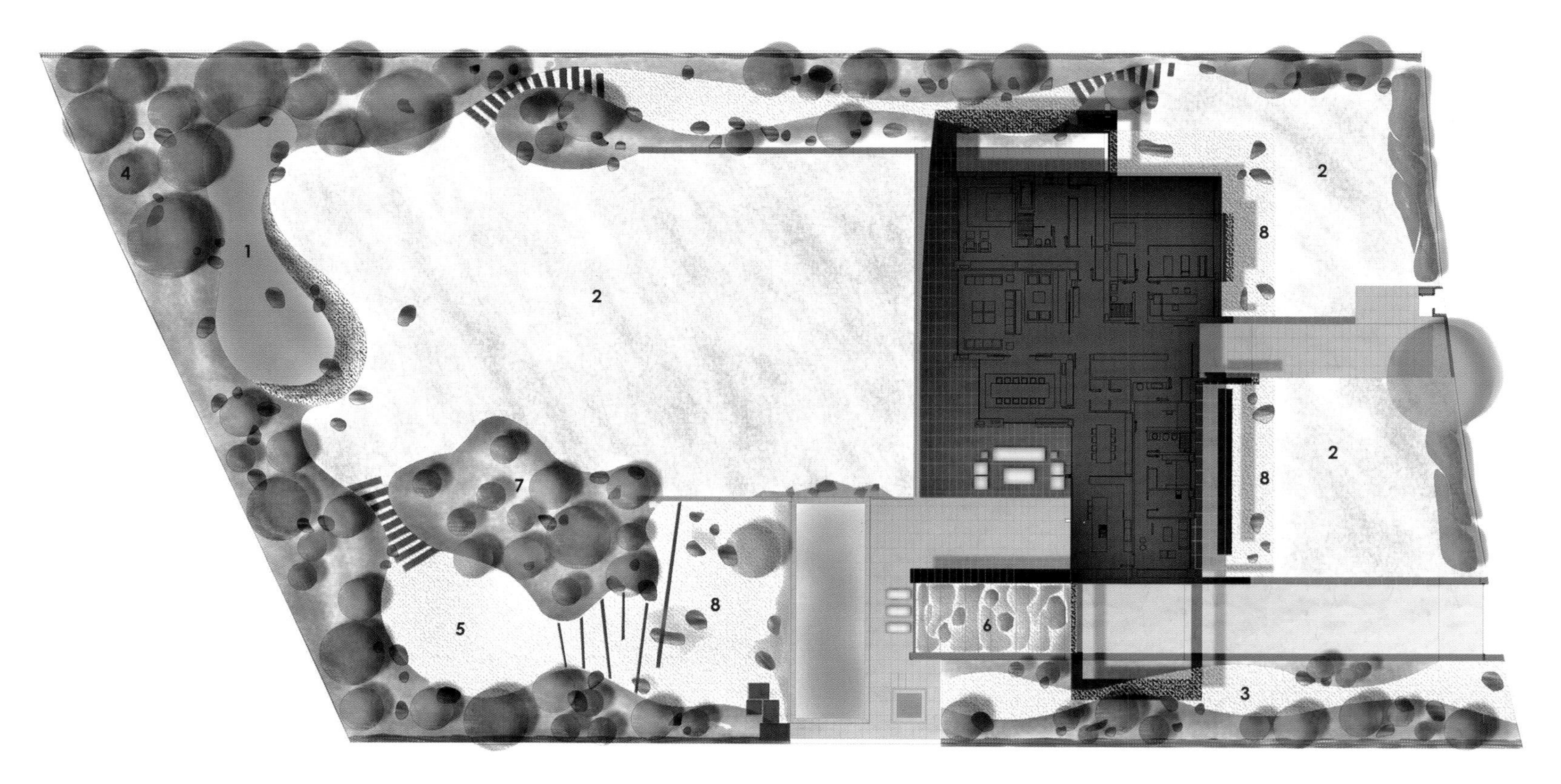

Legend

1.Naturalized lake	2.Lawn	3.Fruit Trees	4.Birch Wood
5.Shaded Area	6.Green Slopes	7.Mixed Woods	8.Zen Garden

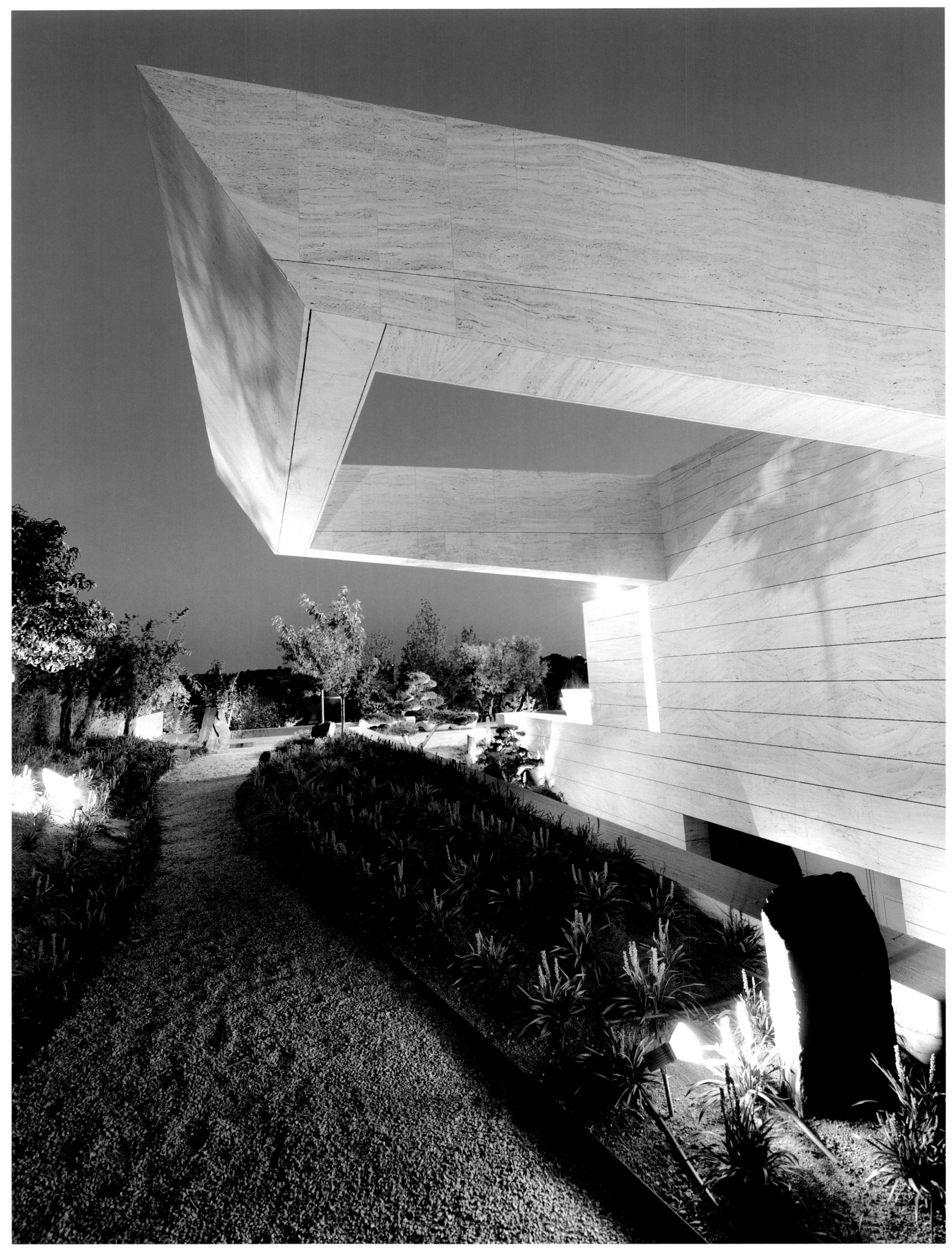

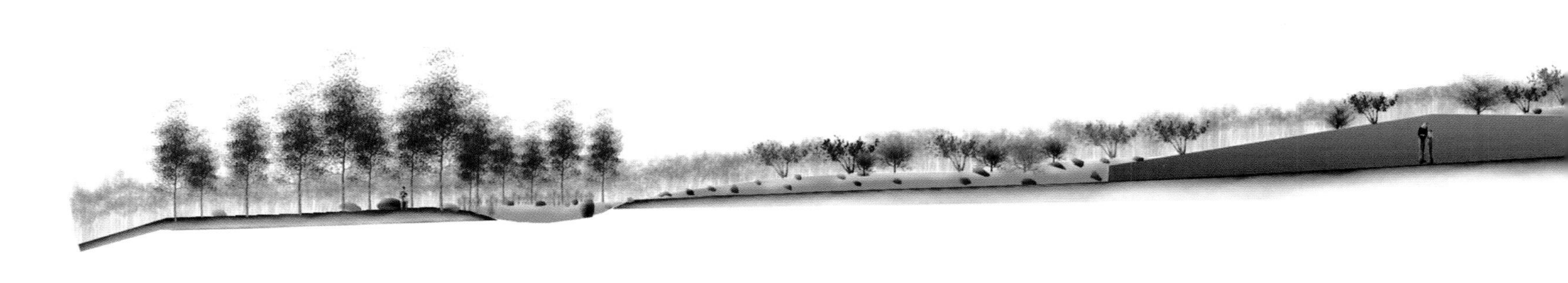

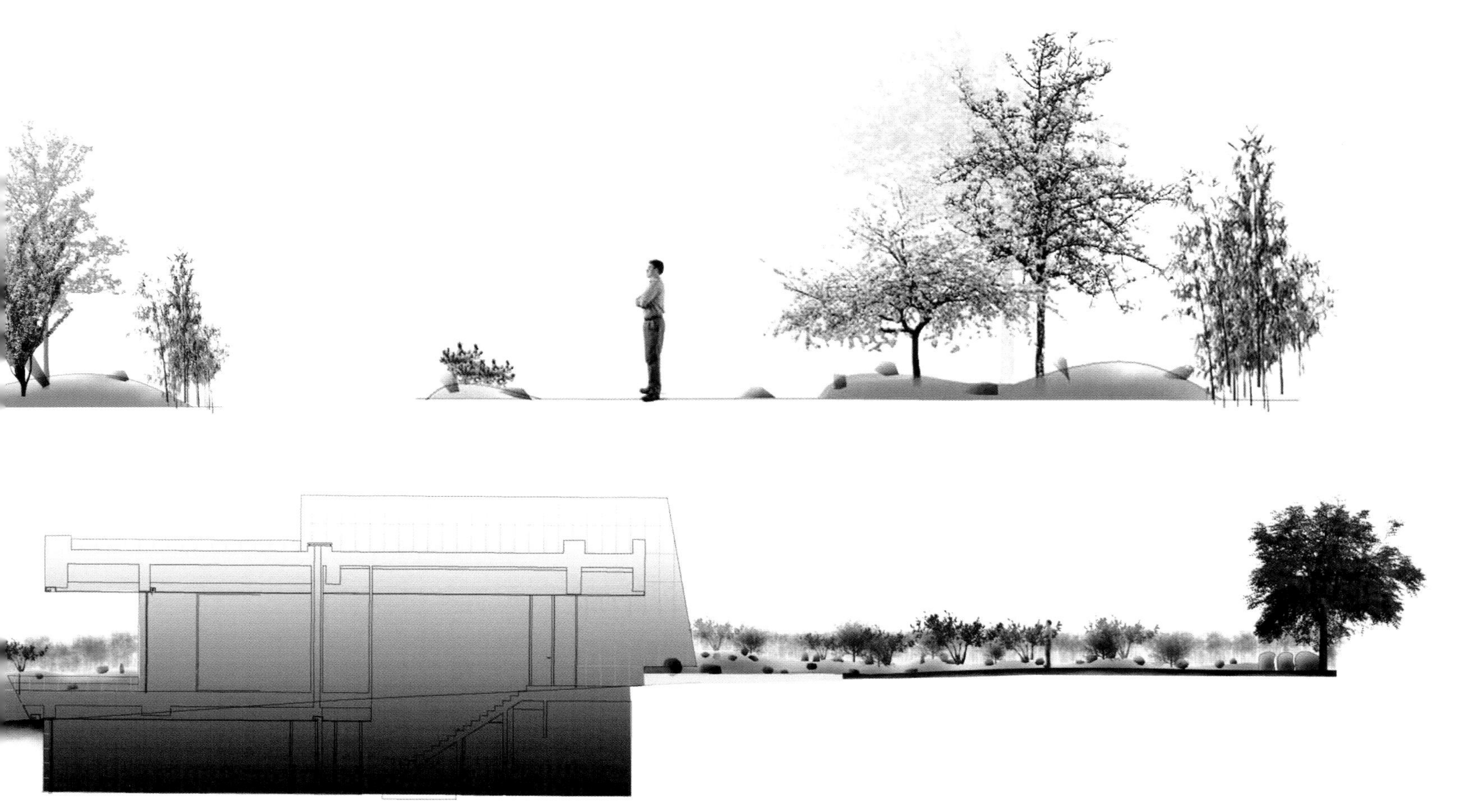

Private Garden nº4 Urbanización Los Lagos Madrid

Landscape Architect
laND30

Architect
A-cero

Location
Madrid, Spain

Area
750 m²

Photographer
A-cero

The landscape architecture project for this villa in 'Urbanización la Finca', Madrid, was made in collaboration with A-Cero architects. The house, built in modern style and clean architecture lines, is contrasted with texturized and organic shaped surroundings as well as sculptural elements while still following the same design language.

The garden is mainly created through land modelling to achieve a series of volumes, different in shape and size, composing the tools that structure and organize the whole space. The different elements originated this way act, according to their shape and size, as hedges, visual barriers between houses, generators of paths, or just as a physical limit.

A large mass of bushes, pruned as cubes, surround small mounds and mark the main entrance. On one side and the back part of the house, occupying the space between the mounds, a large water surface mirrors the sky. While on the other side of the house a sand area fills in the space between mounds offering an appealing place for the young people to play in.

Several cherry trees, comprising the tree layer, are scattered around the garden and distributed through the mounds. The difference in heights generates a rhythm making for an attractive composition.

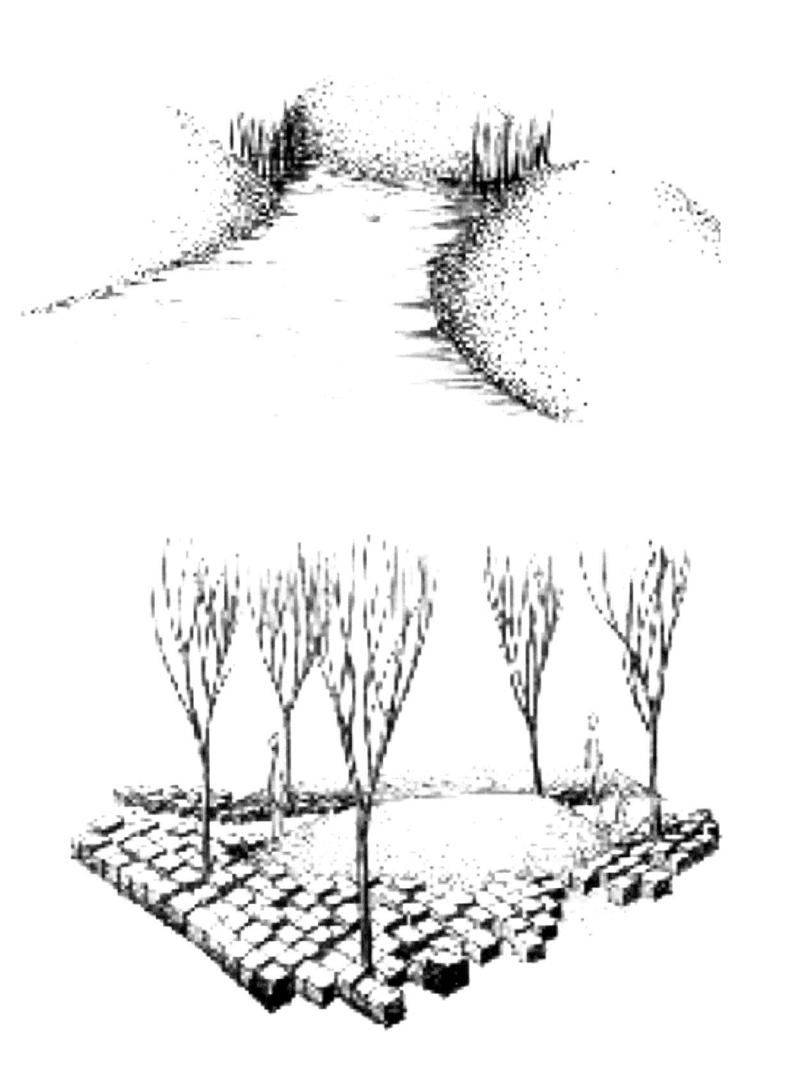

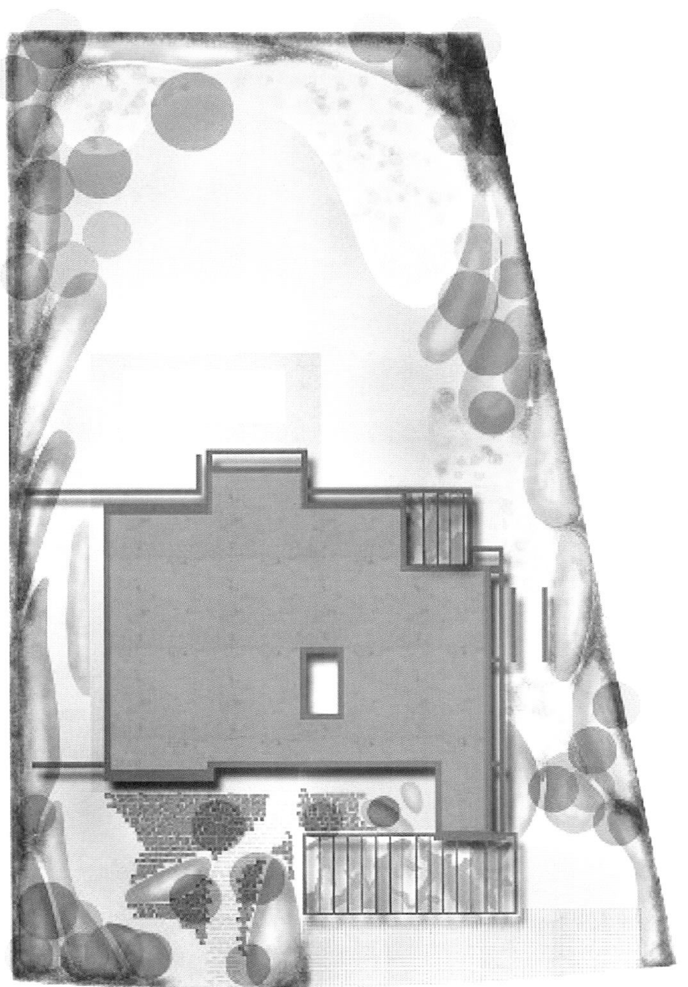

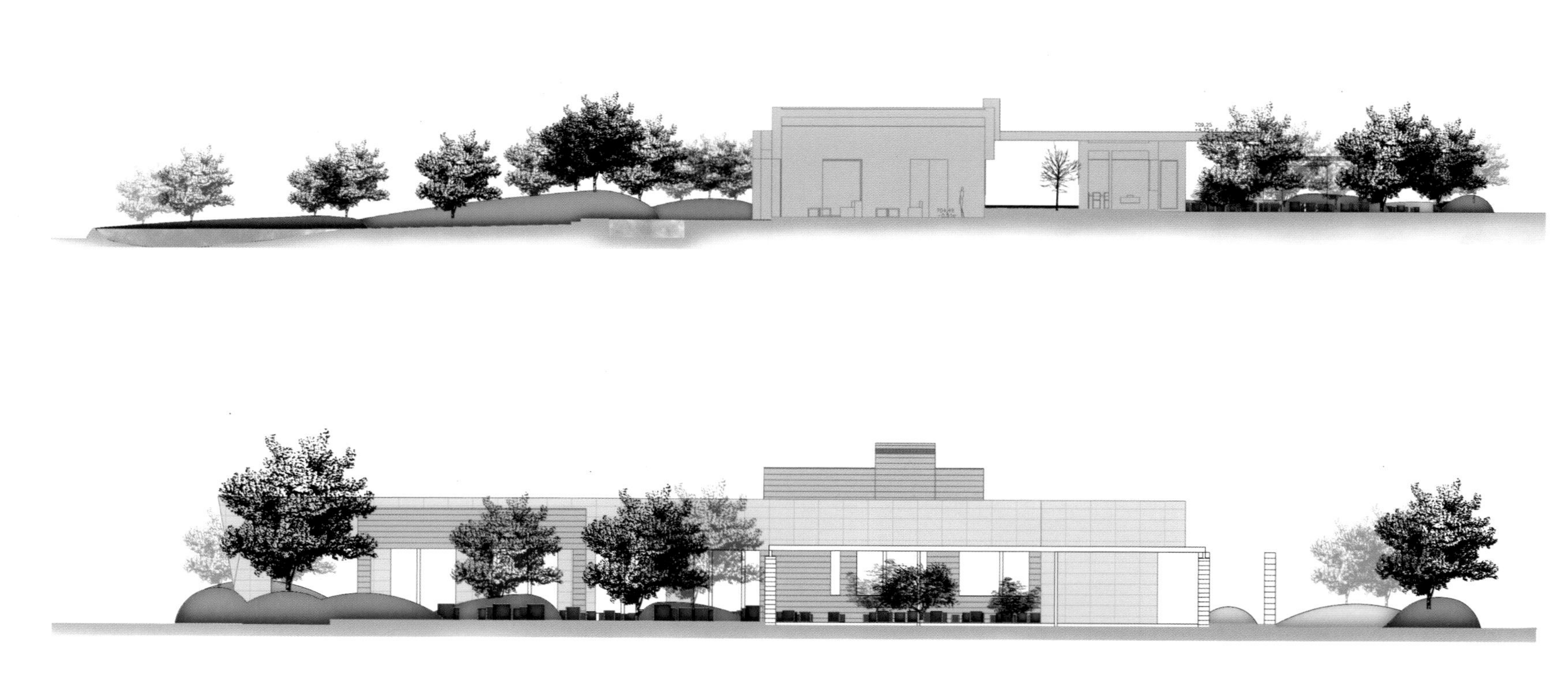

Balaclava Rd

Landscape Architect
Steve Taylor

Firm
COS Design

Builder
David McCallum of DDB Design and Build

Photographer
Urban Angles

The clients wanted to create a slick, clean, timeless outdoor space that allowed for relaxed, informal alfresco entertaining. It needed to be visually striking with a large family pool and a lush yet minimal plant palette and work functionally on a daily basis and for larger numbers when required.

The garden fulfills the brief as stated above. It has a modern, minimalistic Asiatic undertone with the use of the slender weaver bamboo, Japanese maple, cycads and clivia. The clean lines, floating benches, simple palette all combine to create a well-balanced space that will stand the test of time.

The 9 x 4 family pool with feature green Bisazza mosaic bands and raised wall highlighting the slender weaver bamboo create a stunning backdrop to the entire space. Complimented by the floating bench seat to the rear and outdoor shower highlight wall, the pool zone is perfectly balanced to the space. The alfresco roof featuring Aludean laser cut screens creates interest from above and allows valuable light into the area while the floating BBQ Teppan Yaki add functionality to the space in a stylish form.

The front entry highlights a floating plinth with water spilling down both sides of the portico into the lower fish pond. This is a great way to enter the house.

Blackwell Residence

Landscape Architect
Rebecca Dye, Hank Helbush

Firm
Design Focus Landscape Architecture & Construction

Architect
Marty Oakley

Glass Design Artist
Lauren Devon

Owner and General Contractor
Chris Blackwell

Location
Silicon Valley, USA

Photographer
Lauren Devon, Elizabeth Murray

The design challenge for this project was to create an estate garden on a steep landslide site with an existing retaining wall. The designers accomplished this by creating a series of beautiful outdoor rooms at different levels. Since the project was on a landslide site, the design had to allow for reconstruction in the event of more slippage. Because of this, the designers selected imported antique walkway stones from London that were 10.16cm thick and required no mortar or concrete to install.

The designers raised the grade of the pool area by 0.91m to better connect the upper level kitchen and family room with the entertaining area. They also removed a low garage window to create a backdrop for the fireplace and entertaining area.

Because the pool could only be 8.53m long due to landslide restrictions, the designers told the owners it had to look like a jewel to really be significant. They incorporated glass tile on the bottom and along the waterline that could be seen as a parterre view from the home and garden. They also added bronze water sprays from the edges of the pool. The terraced stone stairs and raised beds created a beautiful and dramatic view from the house while increasing usability on the steep site. At the bottom of the stairs they designed a formal flower garden to be changed with the seasons for interest and to provide a focal point.

The bocce ball court and fire pit area were placed at the top of the site to provide another destination with a commanding view. The scope of work included design and site supervision for all elements including master planning, grading, drainage, planting, irrigation, lighting, and masonry.

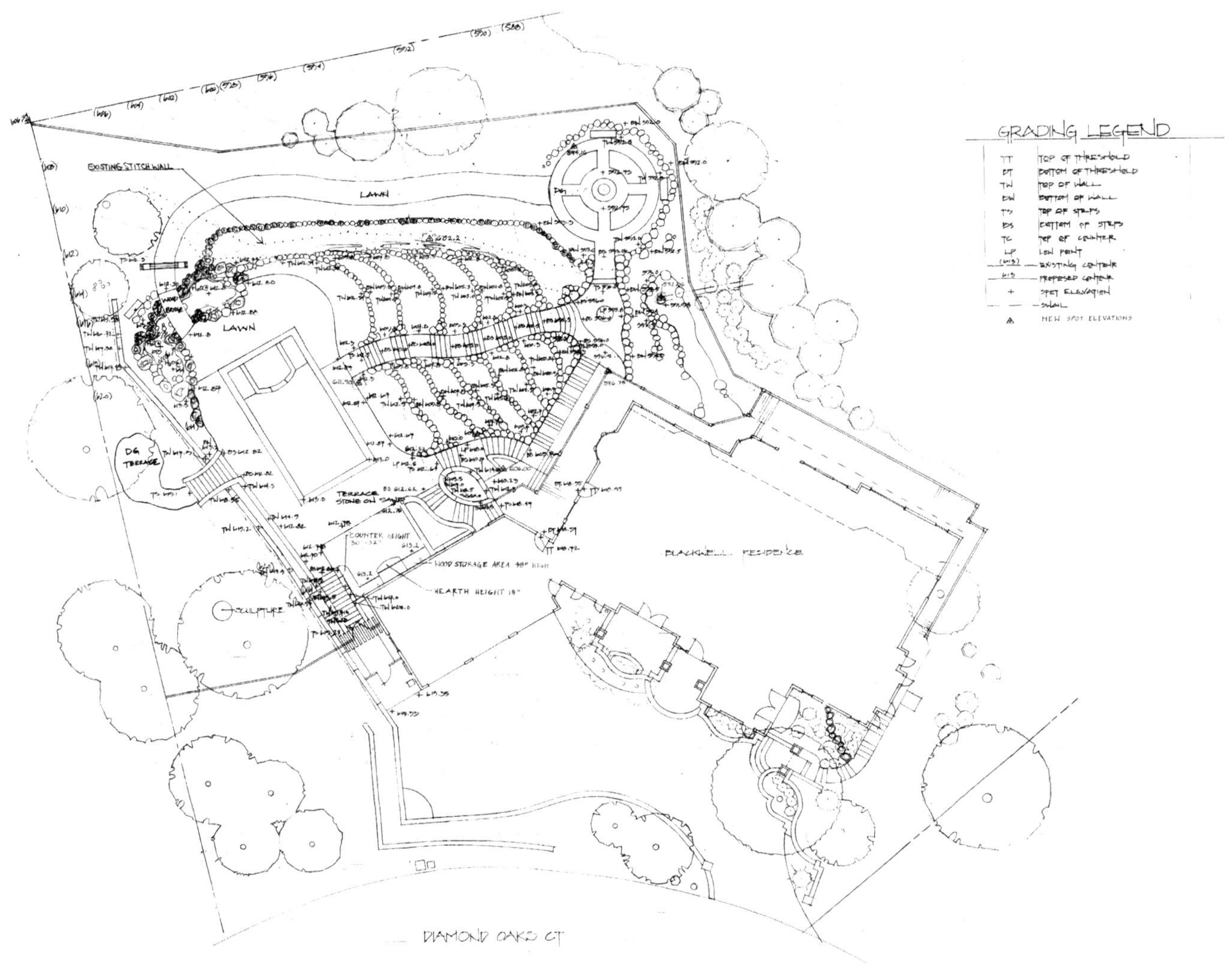
EXISTING STITCH WALL
LAWN
LAWN
DG TERRACE
TERRACE STONE ON SAND
COUNTER HEIGHT
WOOD STORAGE AREA
HEARTH HEIGHT 18"
SCULPTURE
BLACKWELL RESIDENCE
DIAMOND OAKS CT
GRADING LEGEND
TT TOP OF THRESHOLD
BT BOTTOM OF THRESHOLD
TW TOP OF WALL
BW BOTTOM OF WALL
TS TOP OF STEPS
BS BOTTOM OF STEPS
TC TOP OF COUNTER
LP LOW POINT
EXISTING CONTOUR
PROPOSED CONTOUR
SPOT ELEVATION
SWALE
NEW SPOT ELEVATIONS

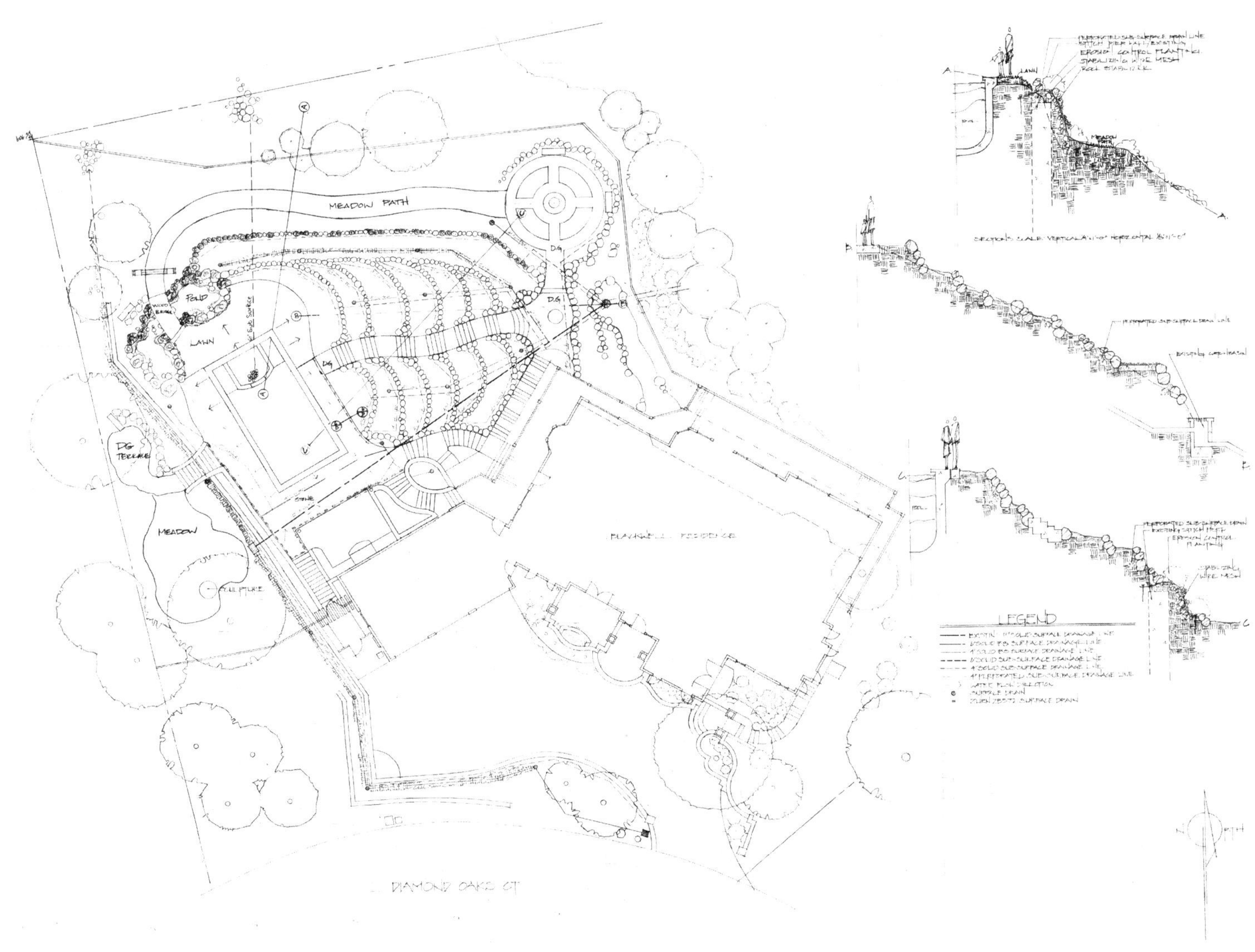
MEADOW PATH
POND
LAWN
MEADOW
DIAMOND OAKS CT
LEGEND

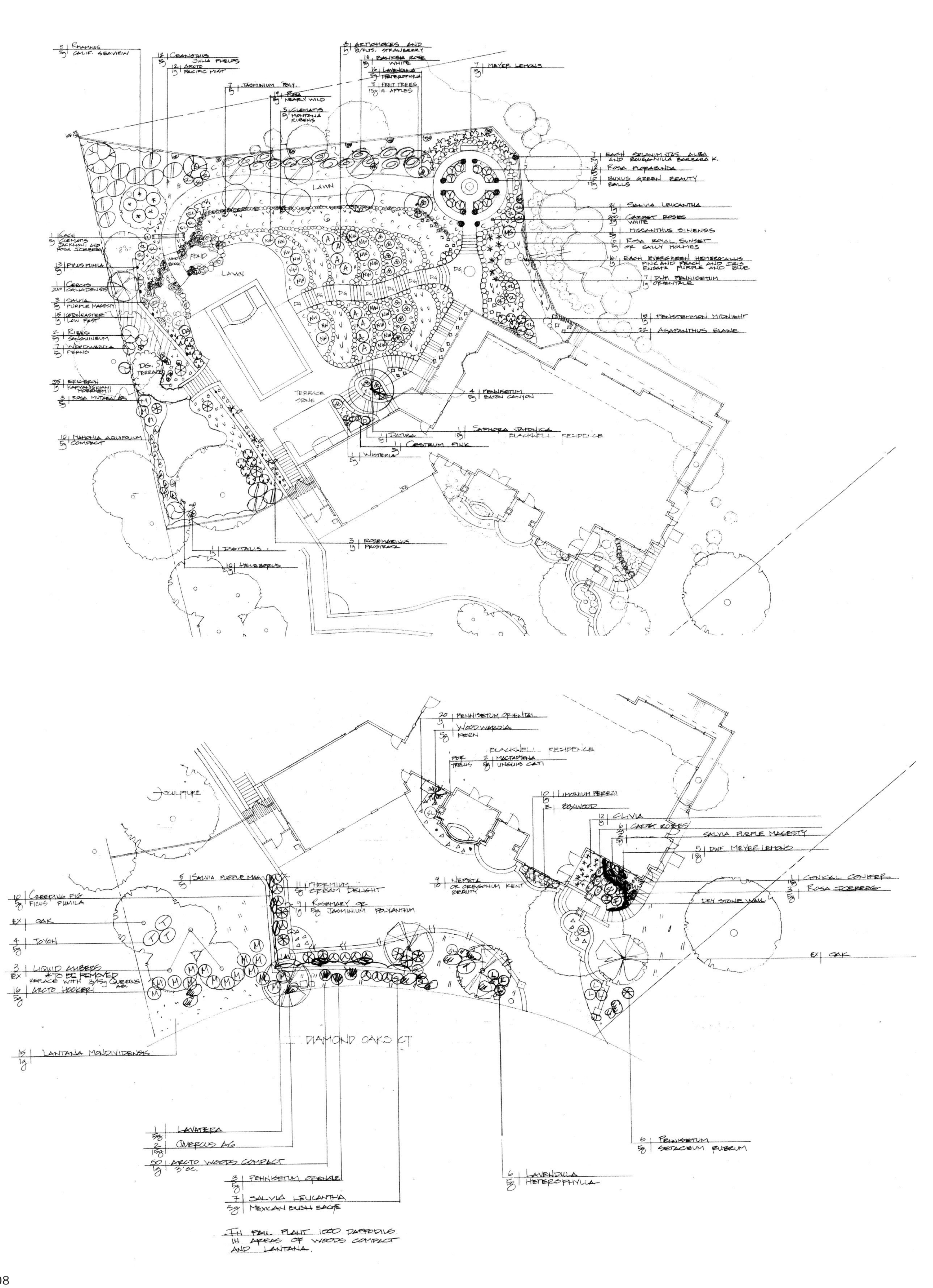

5 5g RHAMNUS CALIF. SEAVIEW
18 5g CEANOTHUS JULIA PHELPS
12 1g ARCTO PACIFIC MIST
7 5g JASMINIUM POLY.
19 1g ROSA NEARLY WILD
5 5g CLEMATIS MONTANA RUBENS
ARTICHOKES AND 8 FLATS STRAWBERRY
16 BANKSIA ROSE WHITE
16 5g LAVENDULA HETEROPHYLLA
8 FRUIT TREES 15g 16 APPLES
7 15g MEYER LEMONS
LAWN
POND
TERRACE STONE
DG TERRACE
1 EACH 5g CLEMATIS JACKMONI AND ROSA ICEBERG
13 FICUS PUMILA
1 24" CERCIS CANADENSIS
3 SALVIA PURPLE MAGESTY
COTONEASTER LOW FAST
2 RIBES SANGUINEUM
7 WOODWARDIA FERNS
35 ERIGERON KARVINSKIANI
3 5g ROSA MUTABILIS
10 5g MAHONIA AQUIFOLIUM COMPACT
7 EACH SOLANUM JAS. ALBA AND BOUGANVILLA BARBARA K.
ROSA FLORABUNDA
BUXUS GREEN BEAUTY BALLS
SALVIA LEUCANTHA
CARPET ROSES WHITE
MISSANTHUS SINENSIS
ROSA ROYAL SUNSET OR SALLY HOLMES
EACH EVERGREEN HEMEROCALLIS PINK AND PEACH AND IRIS ENSATA PURPLE AND BLUE
7 1g DWF. PENNISETUM ORIENTALE
15 PENSTEMMON MIDNIGHT
22 AGAPANTHUS ELAINE
4 5g PENNISETUM EATON CANYON
1 15 SAPHORA JAPONICA
BLACKWELL RESIDENCE
1 5 DATURA
1 5 CESTRUM FINK
1 5 WISTERIA
3 1g ROSEMARINUS PROSTRATA
1 5 DIGITALIS
10 HELEBORUS
20 PENNISETUM ORIENTAL
1 5g WOODWARDIA FERN
BLACKWELL RESIDENCE
FOR TRELLIS 2 5g MACFADYENA UNGUIS CATI
SCULPTURE
10 5 LIMONIUM PEREZII
8 BOXWOOD
12 CLIVIA
6 CARPET ROSES
SALVIA PURPLE MAGESTY
5 15 DWF. MEYER LEMONS
1 15 CONICAL CONIFER
3 5 ROSA ICEBERG
DRY STONE WALL
EX OAK
5 5g SALVIA PURPLE MAG.
11 5g PHORMIUM CREAM DELIGHT
9 5g ROSEMARY OR JASMINIUM POLYANTHUM
9 10 NEPETA OR OREGONUM KENT BEAUTY
10 5g CREEPING FIG FICUS PUMILA
EX OAK
4 5g TOYON
3 EX LIQUID AMBERS *TO BE REMOVED REPLACE WITH 3/15g QUERCUS AG.
16 5g ARCTO HOOKERI
15 1g LANTANA MONDIVIDENSIS
DIAMOND OAKS CT
1 5g LAVATERA
2 15g QUERCUS AG.
50 1g ARCTO WOODS COMPACT 3' OC.
3 5g PENNISETUM ORIENTALE
7 5g SALVIA LEUCANTHA MEXICAN BUSH SAGE
6 5g LAVENDULA HETEROPHYLLA
6 5g PENNISETUM SETACEUM RUBRUM
IN FALL PLANT 1000 DAFFODILS IN AREAS OF WOODS COMPACT AND LANTANA.

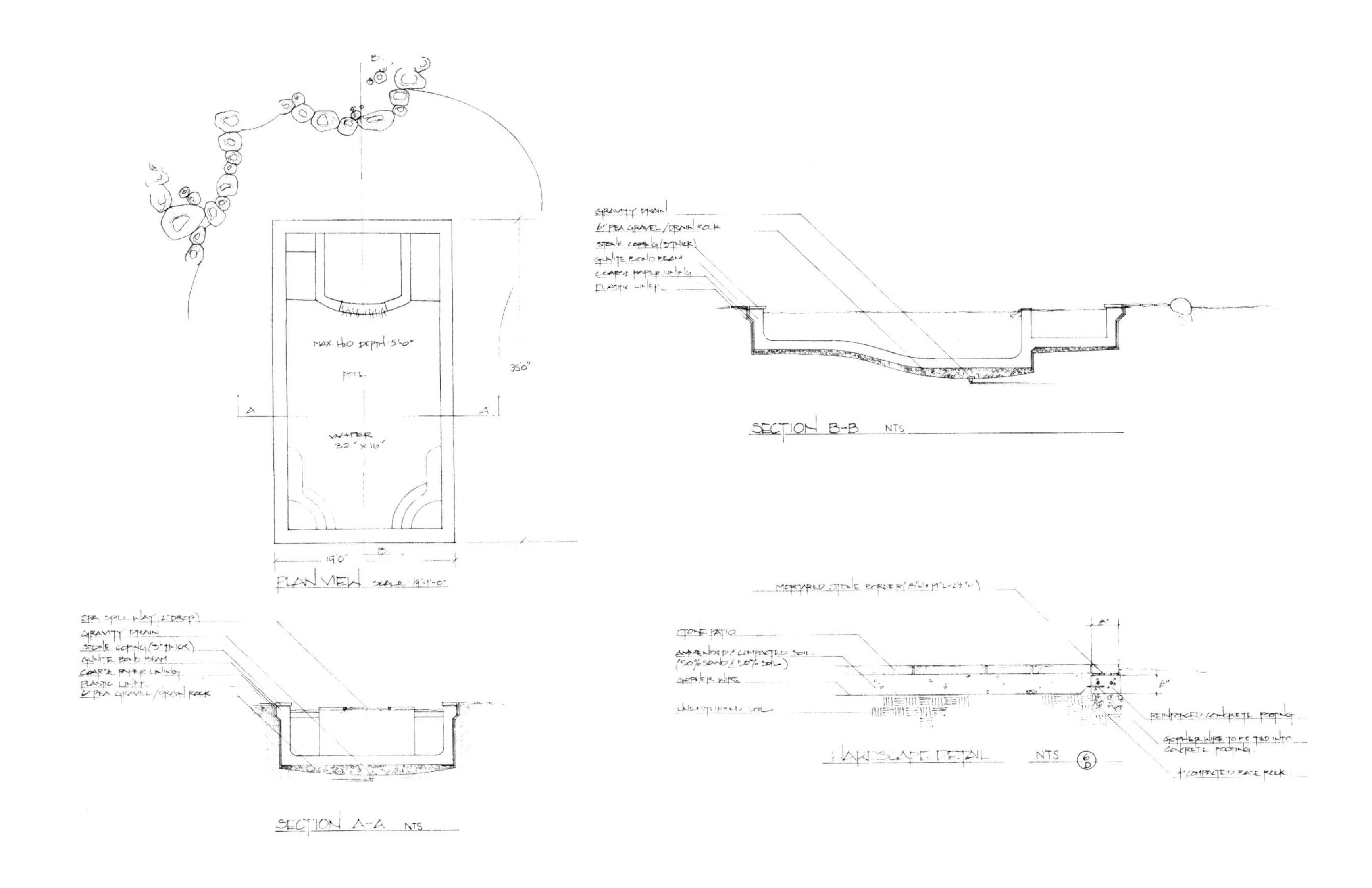
GRAVITY DRAIN
6" PEA GRAVEL / DRAIN ROCK
PLASTIC LINER
SECTION B-B NTS
MAX. H2O DEPTH 5'-0"
POOL
WATER
32' x 16'
350"
19'0"
PLAN VIEW
GRAVITY DRAIN
STONE COPING (3" THICK)
GUNITE BOND BEAM
PLASTIC LINER
6" PEA GRAVEL / DRAIN ROCK
SECTION A-A NTS
STONE PATIO
GOPHER WIRE
REINFORCED CONCRETE FOOTING
HARDSCAPE DETAIL NTS

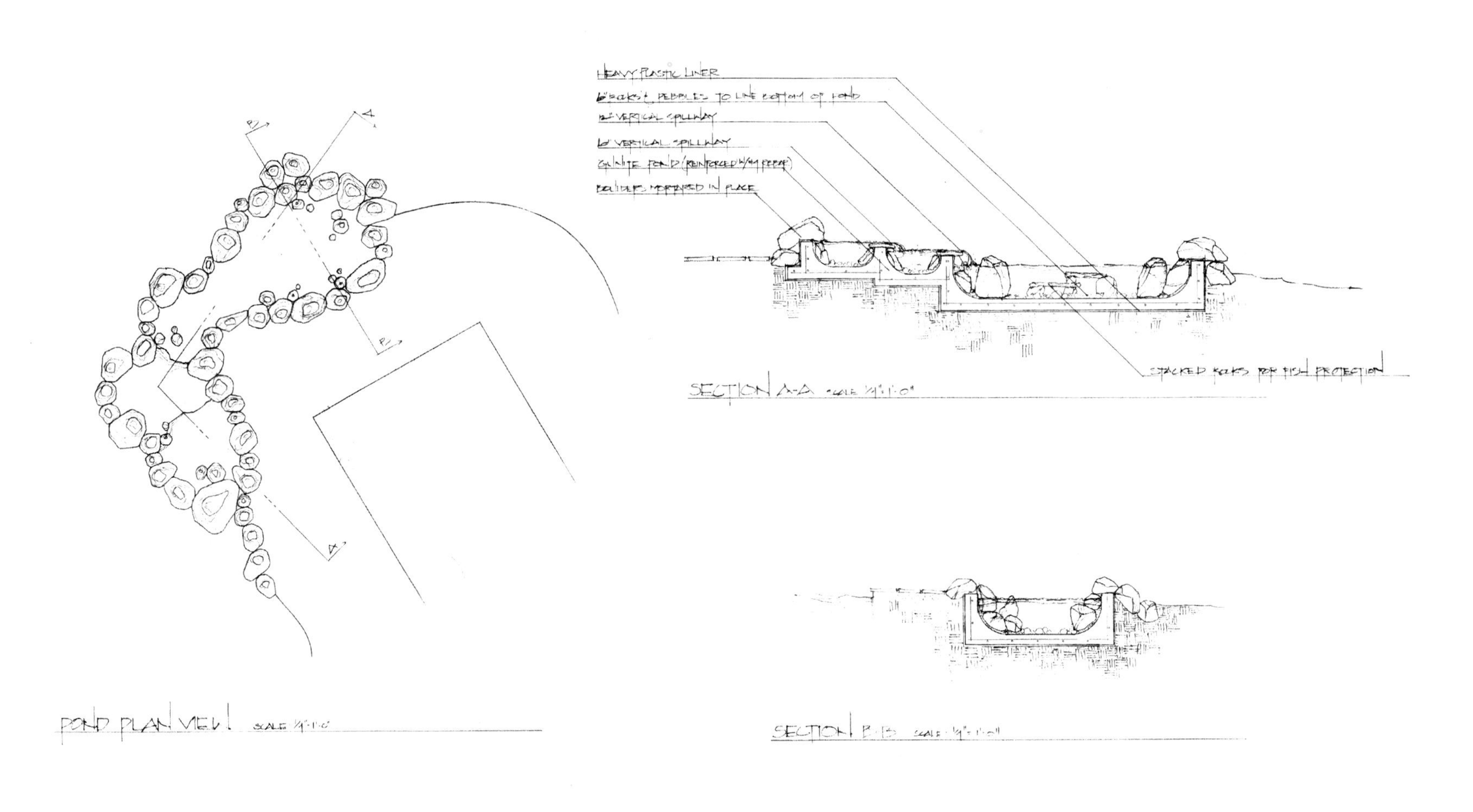
HEAVY PLASTIC LINER
6" VERTICAL SPILLWAY
BOULDERS MORTARED IN PLACE
STACKED ROCKS FOR FISH PROTECTION
SECTION A-A
POND PLAN VIEW
SECTION B-B

Five Realms

Landscape Architect
Marpa Landscape Design Studio

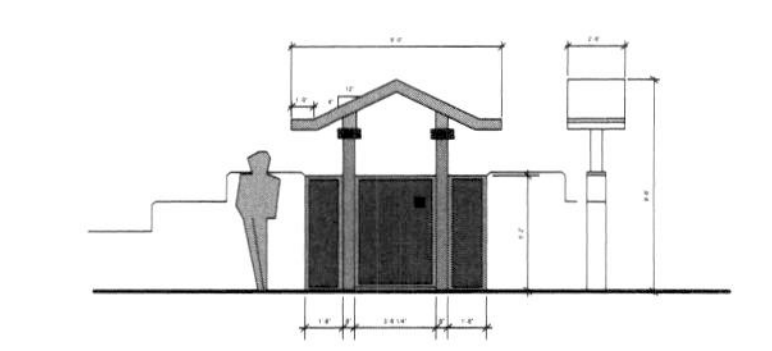

The owners of this energy-efficient country estate are serious meditation practitioners. They asked the designer to create a contemplative environment that was abundant with colour, serene, and exquisitely beautiful. Their plan divided the large space into five realms.

The designers created a boundary for the front garden with a 1.524m retaining wall that provides enough screening to make the entry a lovely surprise. This is the Realm of Enlightenment, where one opens the custom-made front gate, leaving the outer world and entering into a new world. Water originates in a high spring in the north, then cascades into a pool filled with fish and water lilies, and surrounded by colorful alpine plants. In the south is the shrine of the Reclining Buddha. It is made of Cor-ten steel and inlaid with mosaic tile that ranges from pale green through aqua into sky blue, and is under planted with sweet woodruff.

The gardens in the back are prairie-style and seamlessly integrate with the natural surroundings. The large patio looks out onto the Realm of Healing, a five-sector circle of raised planting beds, each with a dominant colour, with an ash tree at the centre. All plants are low in water use.

The Five Realms offer delight to all the senses as well as a quiet, contemplative space for this family.

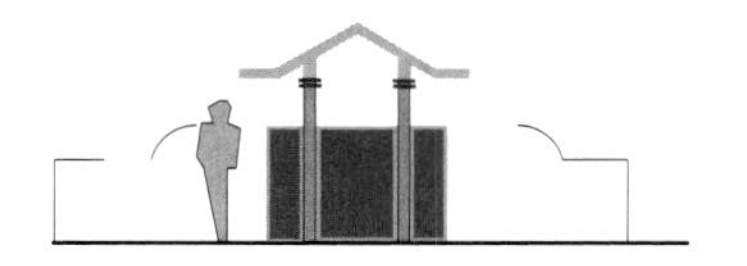

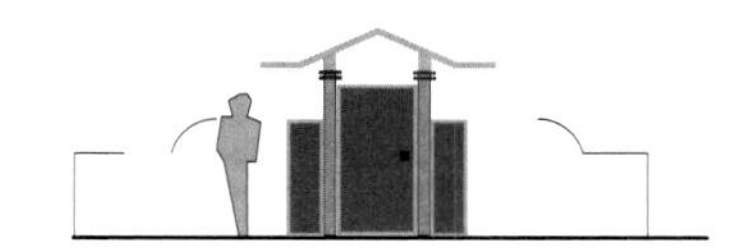

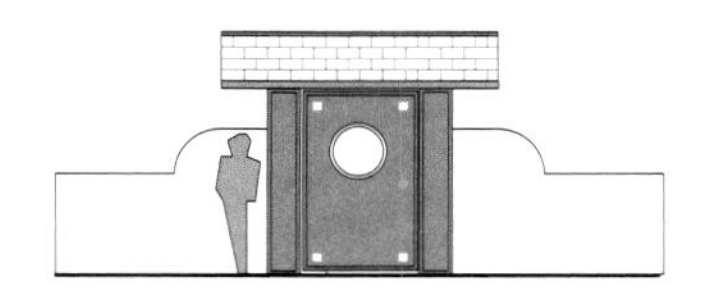

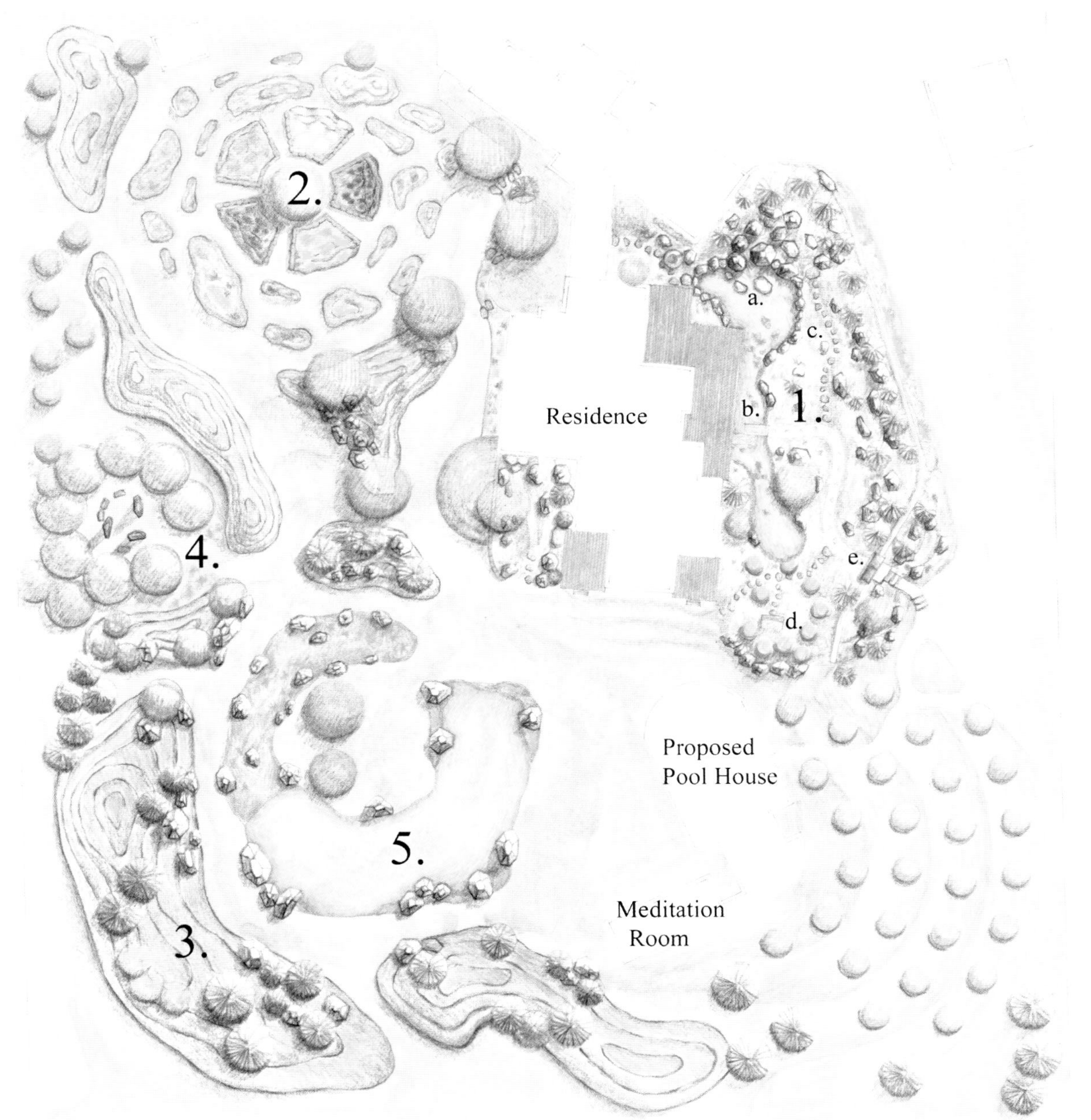

Garden Key

1. Realm of Enlightenment
   - a. Water Source
   - b. Conception Ponds
   - c. Alpine Garden
   - d. Shrine
   - e. Garden Entry
2. Realm of Healing
   Circle of five color flowerbeds
3. Realm of Mountains and Valleys
   Hills reflecting mountains beyond
4. Realm of Mysteries
   Three talking stones surronded by ornamental grasses
5. Realm of Purification
   Continuity of conception ponds emerging as wetlands and naturalized swimming pond

The Five Realms

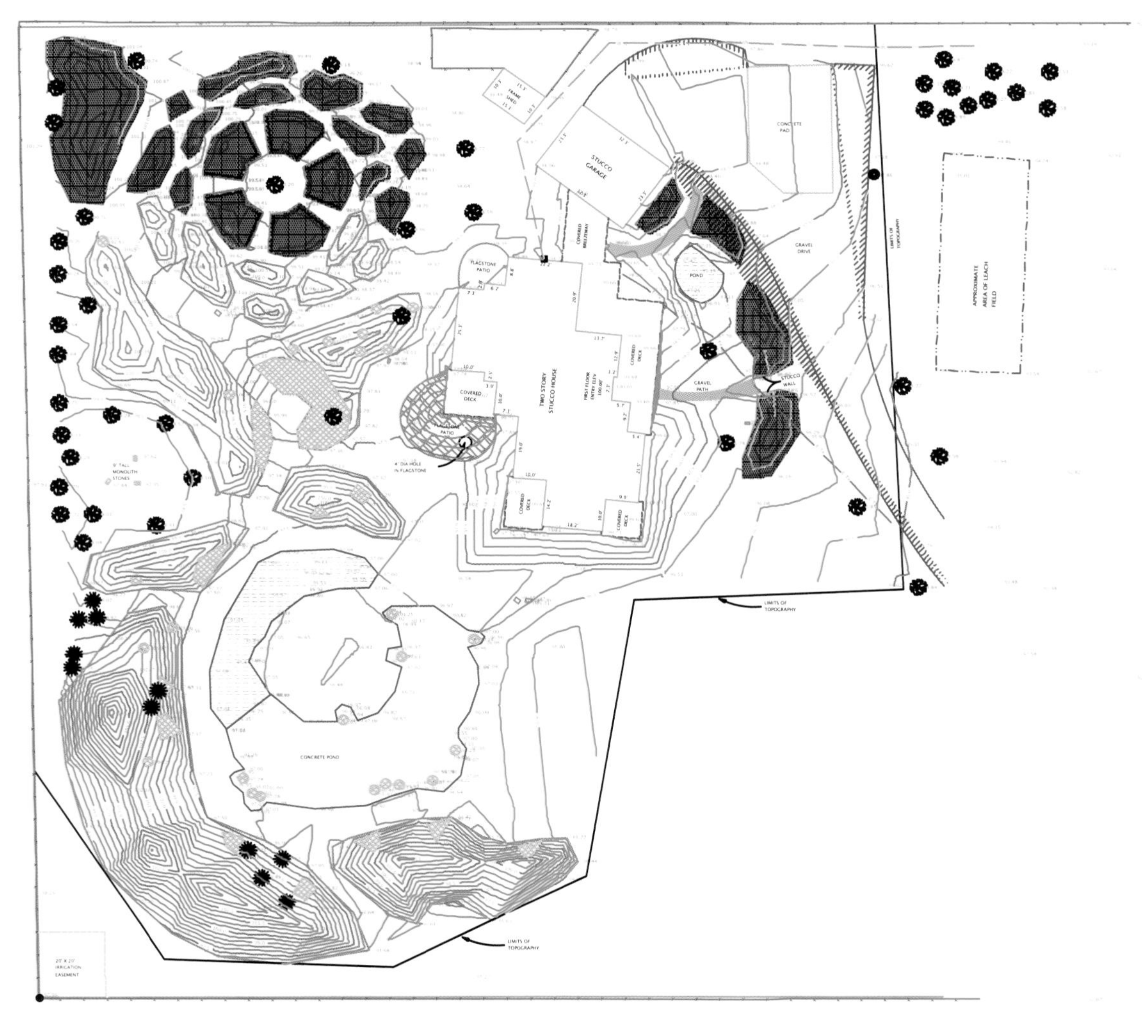

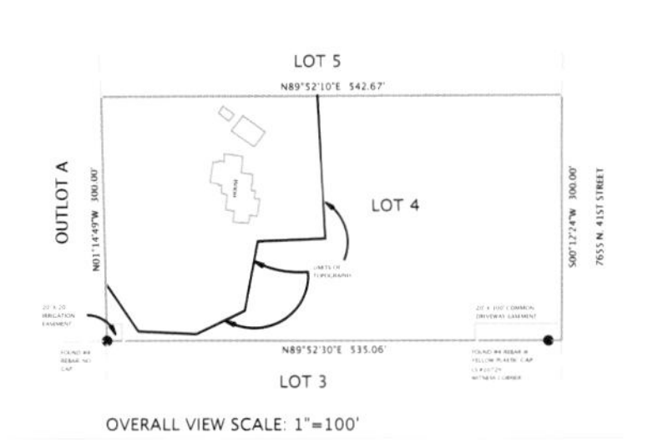

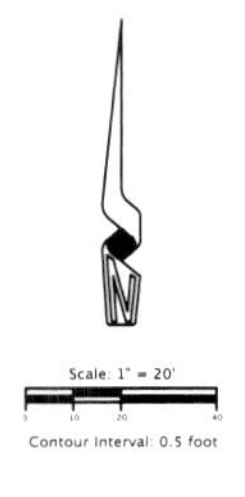

Legend

- FOUND MONUMENT AS DESCRIBED
- GAS METER
- IRRIGATION VALVE BOX
- SEPTIC COVER
- BURIED GAS LINE
- BURIED ELECTRIC/ CABLE TV LINE
- FENCE LINE
- STONE WALL
- ROCK OUTCROP
- LANDSCAPED BERM
- FLAGSTONE
- CONCRETE
- CONIFEROUS TREE
- DECIDUOUS TREE

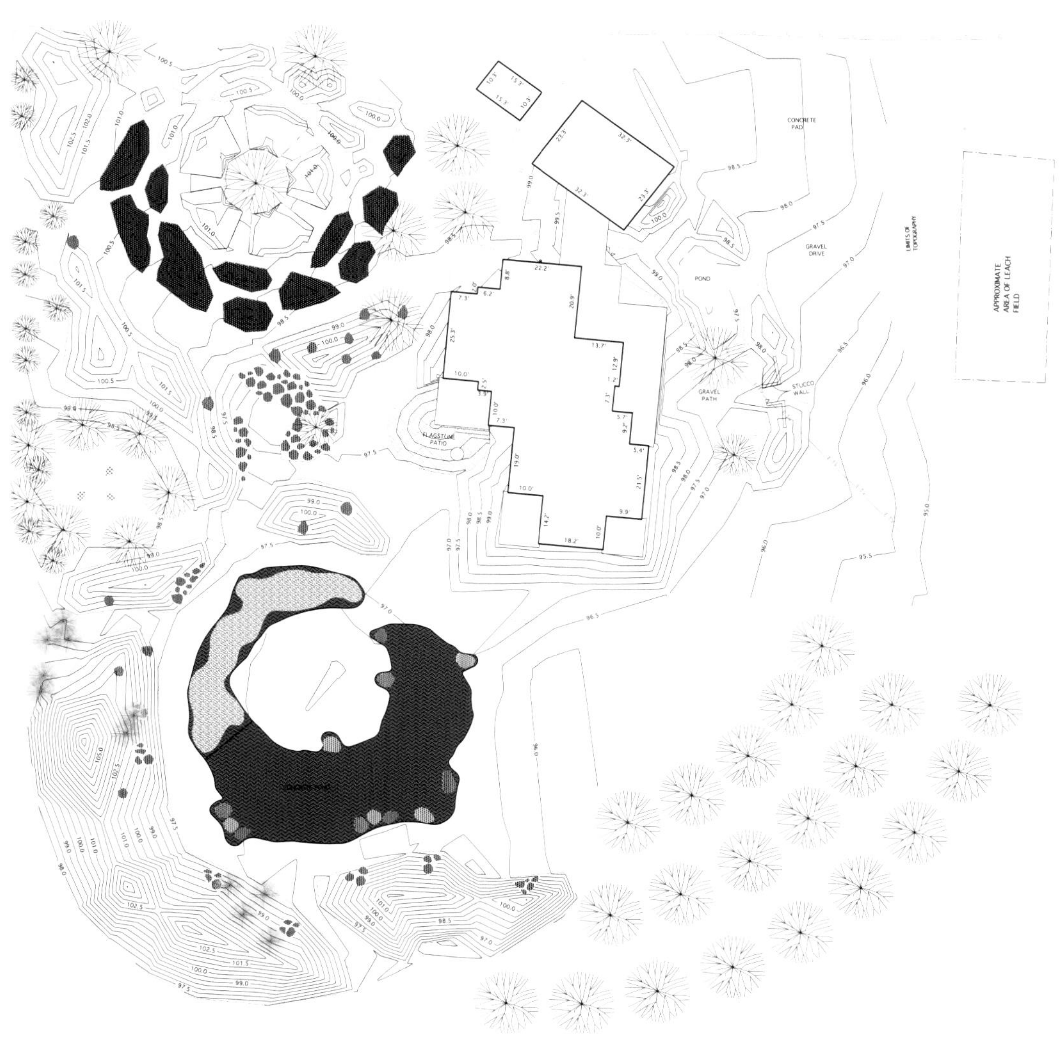
CONCRETE PAD
GRAVEL DRIVE
POND
LIMITS OF TOPOGRAPHY
APPROXIMATE AREA OF LEACH FIELD
GRAVEL PATH
STUCCO WALL
FLAGSTONE PATIO

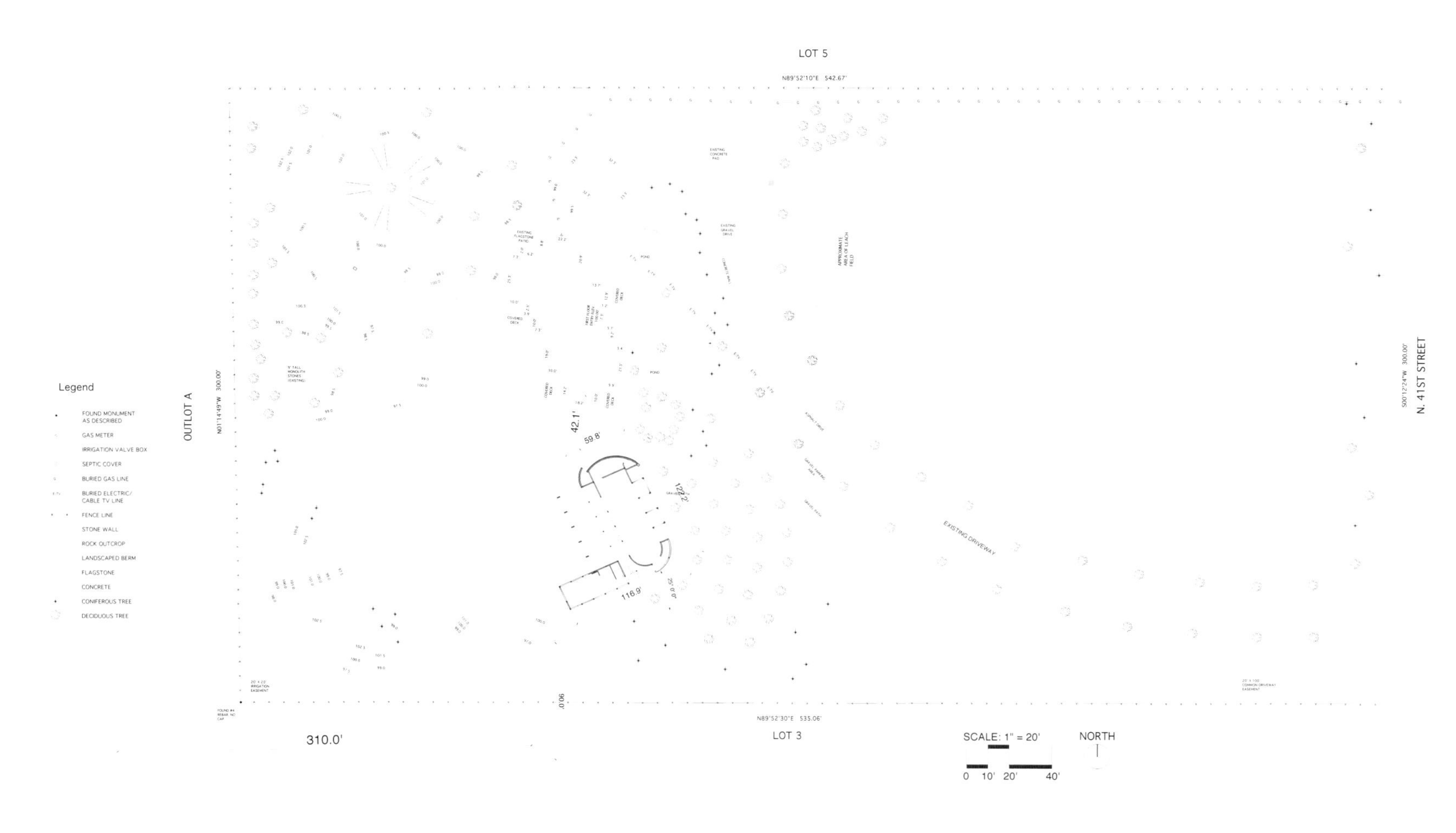

LOT 5
N89°52'10"E 542.67'
Legend
FOUND MONUMENT AS DESCRIBED
GAS METER
IRRIGATION VALVE BOX
SEPTIC COVER
BURIED GAS LINE
BURIED ELECTRIC/ CABLE TV LINE
FENCE LINE
STONE WALL
ROCK OUTCROP
LANDSCAPED BERM
FLAGSTONE
CONCRETE
CONIFEROUS TREE
DECIDUOUS TREE
OUTLOT A
N01°14'49"W 300.00'
42.1'
59.8'
116.9'
EXISTING DRIVEWAY
N. 41ST STREET
310.0'
N89°52'30"E 535.06'
LOT 3
SCALE: 1" = 20'
0 10' 20' 40'
NORTH

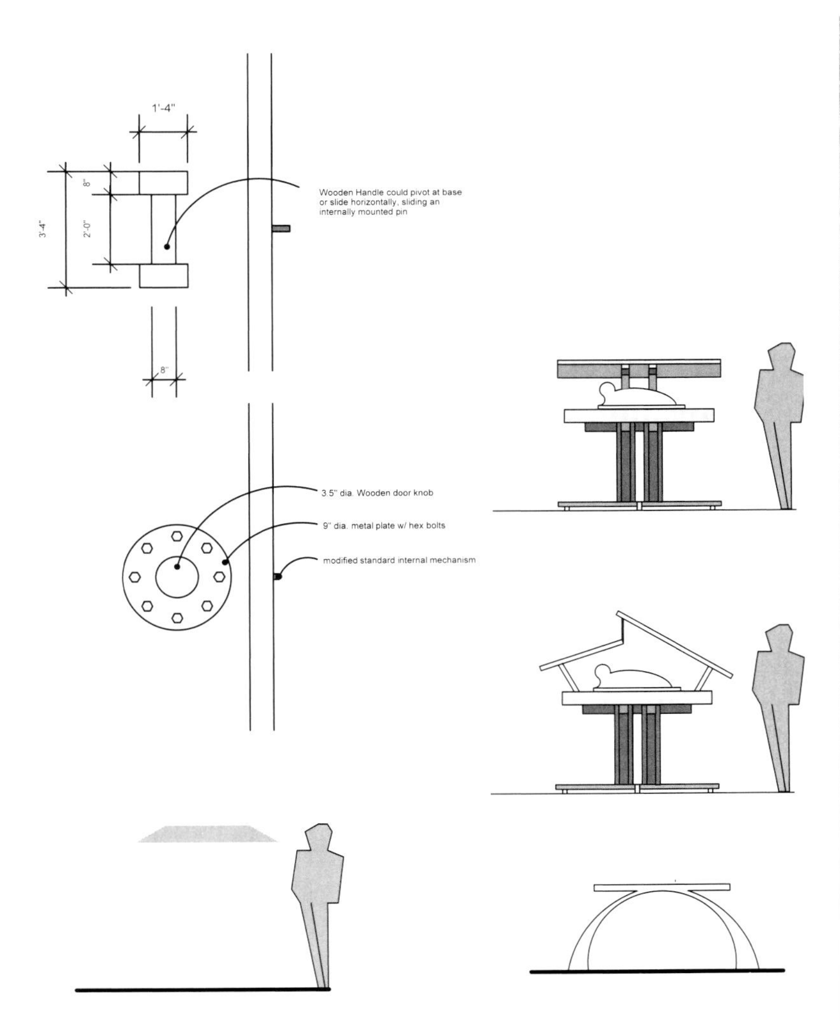
1'-4"
8"
3'-4"
2'-0"
8"
Wooden Handle could pivot at base
or slide horizontally, sliding an
internally mounted pin
3.5" dia. Wooden door knob
9" dia. metal plate w/ hex bolts
modified standard internal mechanism

Harris Residence

Landscape Architect
Rebecca Dye, Hank Helbush,
Nathan Brodie-Rose

Firm
Design Focus Landscape Architecture & Construction

Architect
G&G design and Britt Rowe

General Contractor
Chris Blackwell

Photographer
Lauren Devon

The design intent for this project was to enlarge the small existing terrace and create a major entertaining space that would provide shade and include a pizza oven and barbecue area. The designers used Cameron light stone for the flatwork to give a soft natural feel. The arbor was constructed of steel to afford greater spans and create the curve that mirrored the flatwork shape. Stone columns were designed to ground the arbor and tie it aesthetically to the flatwork.

The existing upper terrace was small and views were cut off by a series of dense balustrades and too many trees. The designers opened the view and created a beautiful new pool and spa that could now be seen and enjoyed from the upper terrace. Adjacent to the pool they created areas for sunbathing surrounded by layered gardens. The grade difference was handled through the use of drystone walls, which also added a feeling of timelessness. By featuring natural materials throughout the project they ensure long lasting environments that only improve with time. They sited the pool house to relate to both the pool and home and designed an intimate arbor and fireplace sitting area. The use of natural materials enriched the entire design of this project.

Several hundred varieties of plants were sensitively placed to afford colour, interest and fragrance throughout the year. A major specimen oak, Quercus lobata, was protected by replacing the existing lawn with a low water meadow. This choice created a luxurious feel and perfect location for a romantic picnic.

EXISTING PLAN

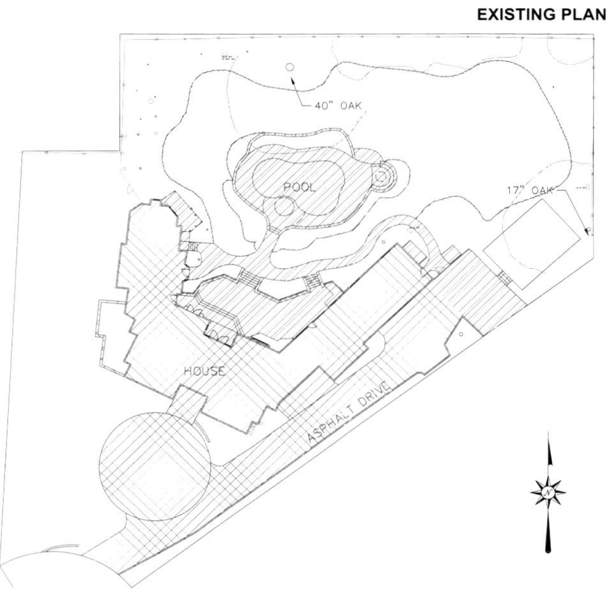

CONCEPT PLAN

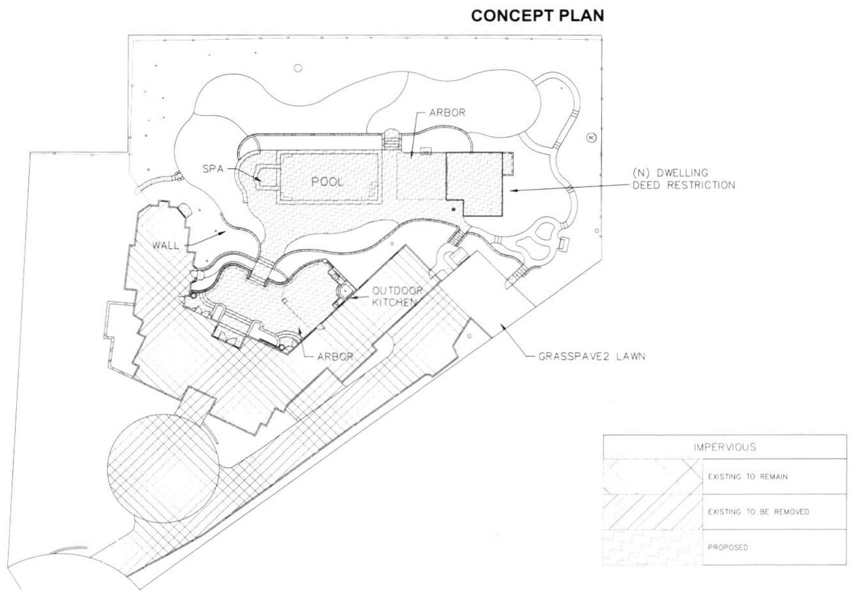

PROJECT INFORMATION

PROJECT DESCRIPTION

LANDSCAPING OF REAR YARD INCLUDING POOL, ARBORS, OUTDOOR KITCHEN, FIREPLACE, HARDSCAPE, IRRIGATION, GRADING, DRAINAGE, & PLANTING

PROJECT DATA

1. ASSESSOR'S PARCEL NUMBER: 397-18-101
2. ADDRESS OF PROJECT:
   14815 THREE OAKS COURT
   SARATOGA, CA 95070
3. OWNER'S NAME: NICK & ELLEN HARRIS
4. EXISTING USE: RESIDENTIAL
5. ZONING DISTRICT: R-14,000
6. SIZE OF LOT: 40,434 SQ. FT.

CONTACT INFORMATION

LANDSCAPE ARCHITECT

DESIGN FOCUS
Rebecca Dye
PO Box 485
Ben Lomond, CA 95005
Tel: (831) 336-3100
Fax: (831) 336-3700
Email:rjd@designfocus.com

NOTE: ENGINEERING SHEETS S1, S2, & S3 TAKE PRECEDENT IN ANY DISCREPANCIES IN THESE DOCUMENTS.

TABLE OF CONTENTS

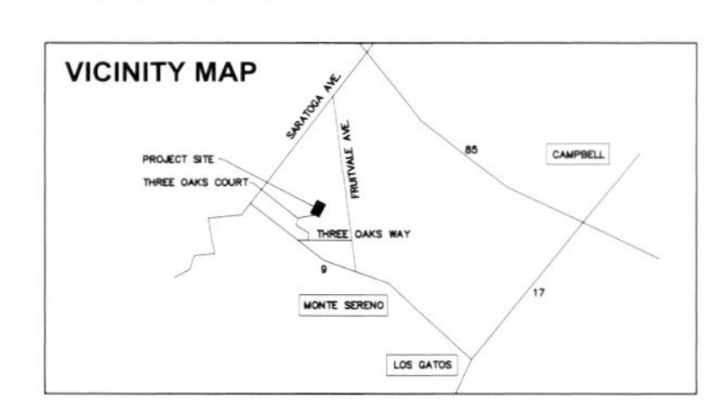

IMPERVIOUS CALCULATIONS

AREA		sq ft	%
Allowable Impervious Coverage		**15,567**	**38%**
Size of Lot (gross)		40,434	100%
Existing Impervious Coverage to Remain			
House		4,337	
Hardscape & Driveway (to remain)		5,069	
Shed		34	
	Total	9,440	23%
Existing Impervious to be Removed			
Hardscape & Concrete Band		4,456	
Driveway to be Grasspave (sport court)		732	
	Total	5,188	13%
Proposed Impervious (New)			
Hardscape, pool, pool house, walls, shed, railroad tie steps		5,579	
Balcony		21	
	Total	5,600	14%
Summary			
Existing Impervious Coverage		**14,628**	**36%**
Existing Impervious to be Removed		**5,188**	**13%**
Proposed Impervious (New)		**5,600**	**14%**
	Total Proposed Impervious for site:	**15,040**	**37%**

Japanese Jewel

Landscape Architect
Ian Smith

Firm
Acres Wild

Location
Buckinghamshire, England

Area
12,140 m^2

This Japanese style garden is a small, enclosed garden space within a much larger country garden in rural Buckinghamshire. The garden was originally laid to lawn with some existing trees, including a beautiful Dove or Handkerchief Tree (Davidia).

The clients wanted a tranquil and contemplative Japanese style garden to compliment a new pool house, and to provide an attractive garden on the way to the swimming pool and tennis court.

Square black basalt paving slabs are used to create a terrace off the pool house deck and they fragment out into a curvilinear shaped lawn. Two granite 'bridges' cross a gravel stream and granite stepping stones create informal paths to connect the garden's entry points to the lawn. Stone elements like a bench and oriental lanterns create focal points in the garden, and trims, steps and details are reminiscent of Japanese Gardens. The planted borders also contribute to the oriental character and atmosphere and they wrap around the existing trees and contain the lawn, terrace and stream. Nearly all of the plants used in the garden are native to Japan, including bamboos for screening, maples and pines for feature plants, and Pachysandra and Liriope for evergreen groundcover.

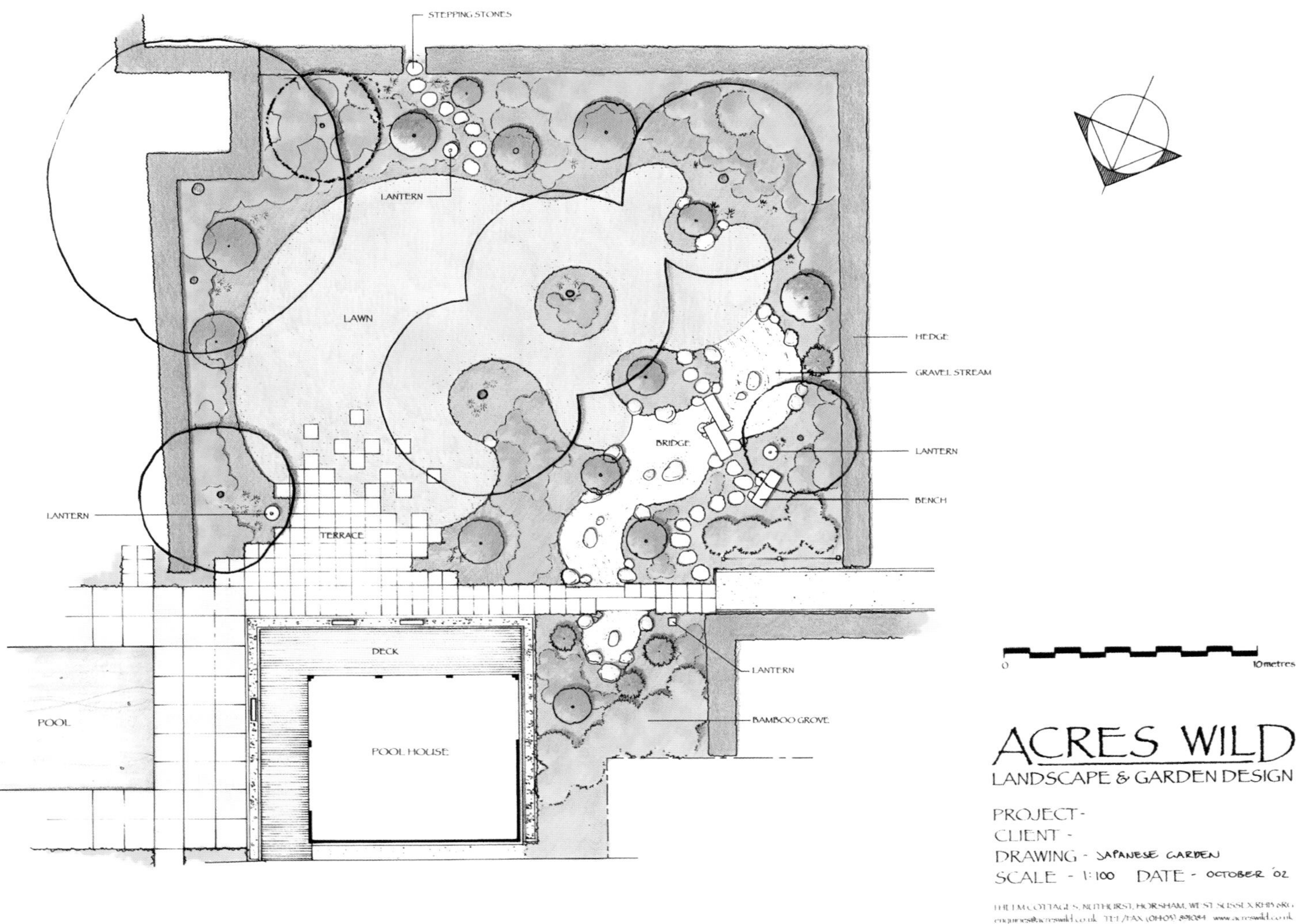
STEPPING STONES
LANTERN
LAWN
HEDGE
GRAVEL STREAM
BRIDGE
LANTERN
BENCH
LANTERN
TERRACE
DECK
LANTERN
POOL
POOL HOUSE
BAMBOO GROVE
0
10metres
ACRES WILD
LANDSCAPE & GARDEN DESIGN
PROJECT-
CLIENT -
DRAWING - JAPANESE GARDEN
SCALE - 1:100 DATE - OCTOBER '02

Rose Garden Residence

Landscape Architect
Rebecca Dye, Hank Helbush

Firm
Design Focus Landscape Architecture & Construction

Architect
Mosher Associates Architecture

Location
Silicon Valley, USA

Photographer
Lauren Devon, Elizabeth Murray

The client asked the designers to design a white garden with French influences that would relate to the architecture of the original 1920's home.

The entry garden was designed with detailed brick walls and iron gates, which leads visitors to the antique granite entry path and courtyard. The granite stones were imported from China, and are several hundred years old. A large pond filled with white water lilies and shimmering gold fish served as the centrepiece of the courtyard. For the centre of the pond the designers found a large antique urn and filled it with seasonal flowers.

Walking past a classic French iron gazebo, the path leads to the library garden, a quiet space for relaxation and reflection. For this area they included several pieces of antique sculpture as well as a French limestone wall fountain.

The main bluestone terrace was designed to serve large groups while still being in scale with the home and creating a sense of intimacy through elements such as the spectacular outdoor fireplace, antique urns and sculpture. The central sycamore adds intimacy by dividing up spaces for different seating areas. Iceberg roses and lavender surround the main terrace, and Sally Holmes roses climb 3.66m up the walls giving an old world feel to the space.

In addition to the roses, lavender and other more formal plants, the designers incorporated a variety of plants to encourage habitat. Since completion of the project, there have been many nesting birds on the site and the owner has connected with the nature in her garden.

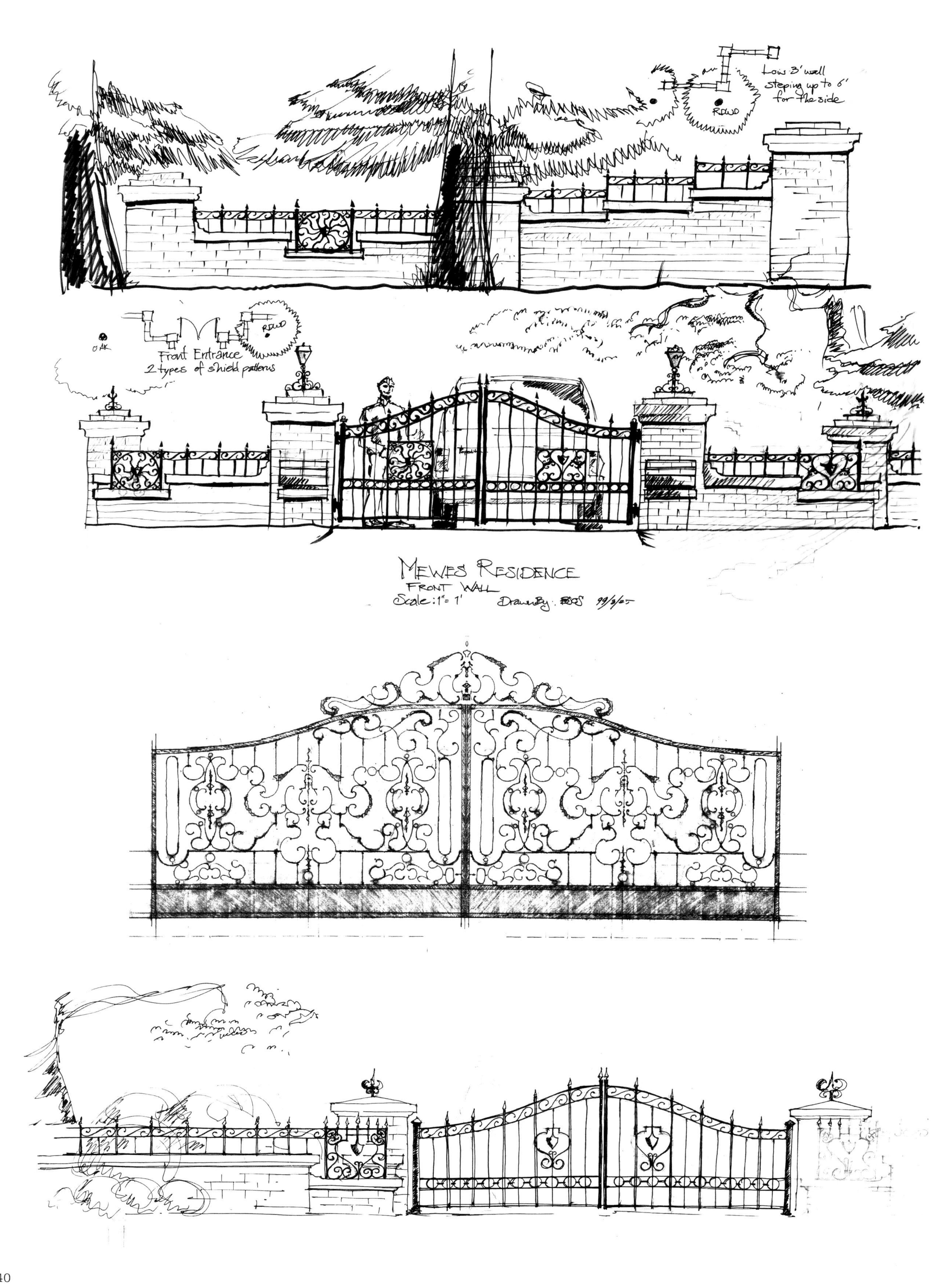
Low 3' wall
steping up to 6'
for the side
RDWD
OAK
Front Entrance
2 types of shield patterns
RDWD
MEWES RESIDENCE
FRONT WALL
Scale: 1" = 1'
Drawn By:
99/3/05

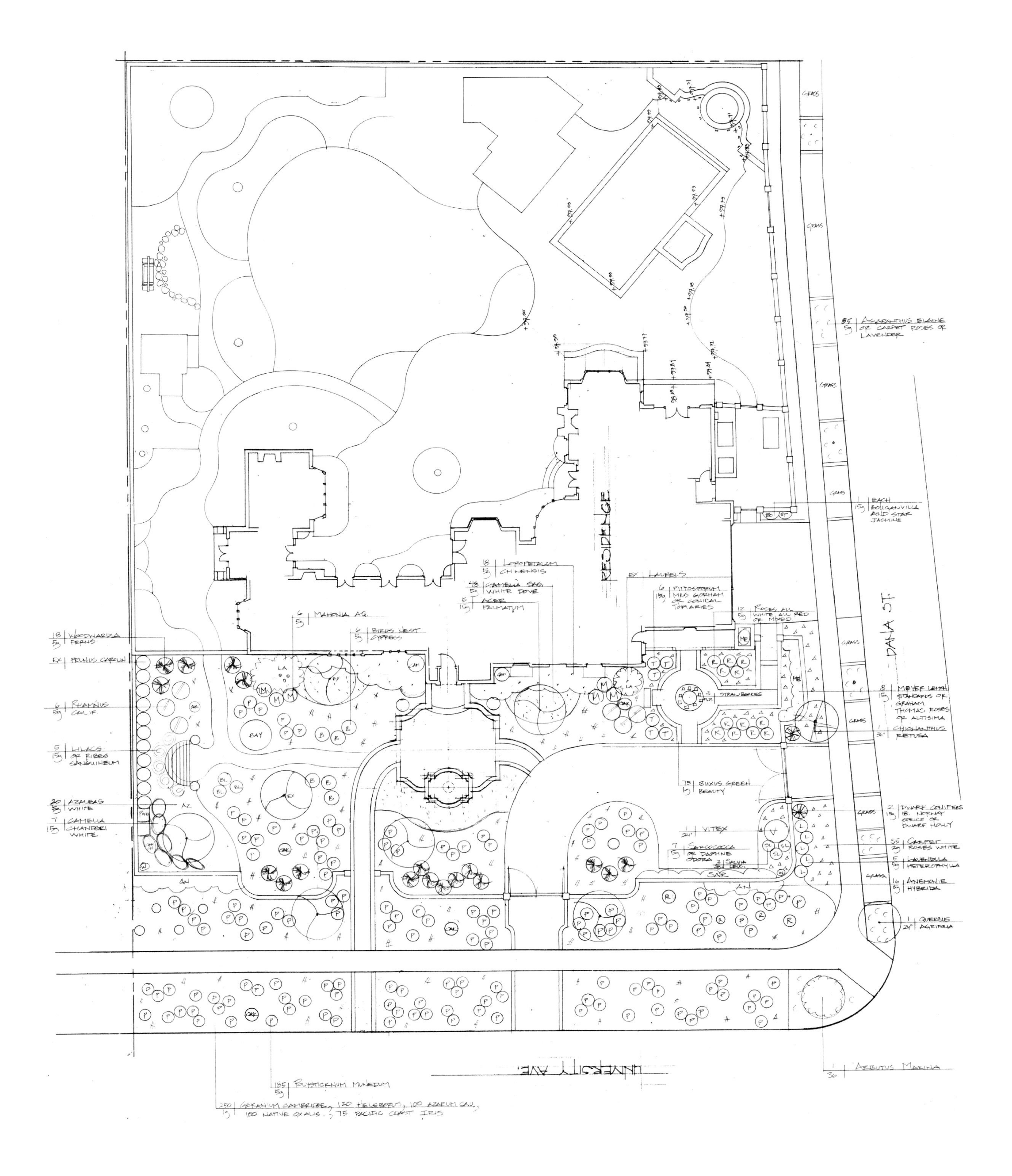
RESIDENCE
18 | WOODWARDIA 5g | FERNS
EX | PRUNUS CAROLIN
6 | RHAMNUS 5g | CALIF
5 | LILACS 15g | OR RIBES SANGUINEUM
20 | AZALEAS 5g | WHITE
7 | CAMELLIA 15g | CHANDERI WHITE
6 | MAHONIA AQ. 5g
6 | BIRDS NEST 5g | CYPRESS
18 | LOROPETALUM 5g | CHINENSIS
48 | CAMELLIA SAS. 5g | WHITE DOVE
5 | ACER 15g | PALMATUM
EX | LAURELS
6 | PITTOSPORUM 15g | MRS GORHAM OR CONICAL TOPIARIES
12 | ROSES ALL 5g | WHITE ALL RED OR MIXED.
3 | STRAWBERRIES
85 | AGAPANTHUS ELAINE 5g | OR CARPET ROSES OR LAVENDER
GRASS
1 | EACH 15g | BOUGANVILLA AND STAR JASMINE
DAHLA ST.
8 | MEYER LEMON 15g | STANDARDS OR GRAHAM THOMAS ROSES OR ALTISIMA
1 | CHIONANTHUS 36" | RETUSA
75 | BUXUS GREEN 1g | BEAUTY
1 | VITEX 24"
7 | SARCOCOCCA 5g | OR DAPHNE ODORA
3 | SALVIA
2 | DWARF CONIFERS 15g | IE. NORWAY SPRUCE OR DWARF HOLLY
55 | CARPET 2g | ROSES WHITE
5 | LAVENDULA 5g | HETEROPHYLLA
16 | ANEMONIE 5g | HYBRIDA
1 | QUERCUS 24" | AGRIFOLIA
UNIVERSITY AVE.
1 | ARBUTUS MARINA 36
185 | POLYSTICHUM MUNITUM 5g
250 | GERANIUM CAMBRIDGE, 120 HELLEBORUS, 100 AZARUM CAU., 1g | 100 NATIVE OXALIS, 75 PACIFIC COAST IRIS

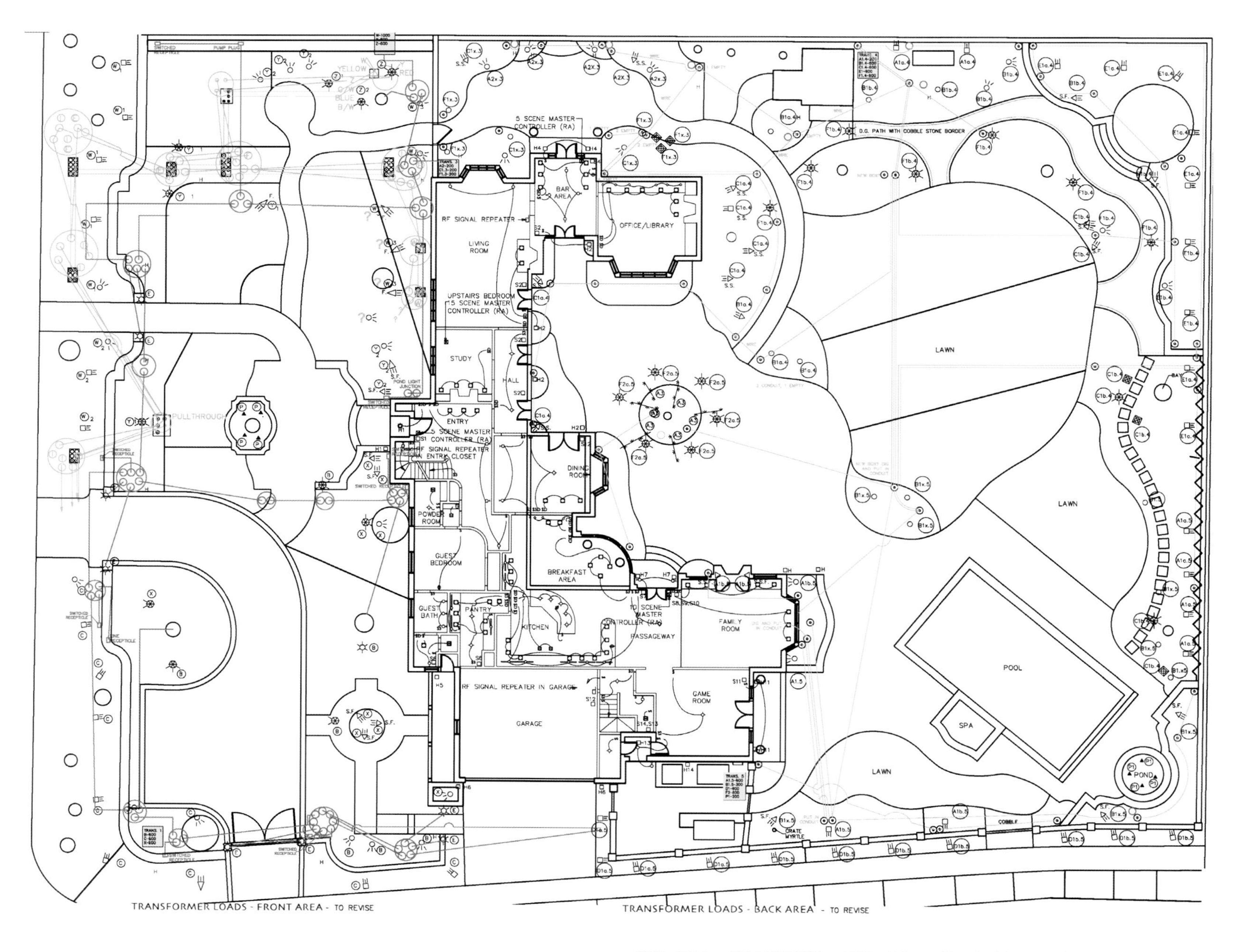

LEGEND

- TREE MOUNTED BULLET SPOT DOWNLIGHT
- STRING LIGHTS
- ROUND WALL FLOOD
- SQUARE WALL FLOOD
- F. BULLET FLOOD
- S. BULLET SPOT
- S.F. SMALL BULLET FLOOD
- S.S. SMALL BULLET SPOT
- WALL LIGHTS
- COLUMN LIGHTS
- S.S.F. MR7 BULLET UPLIGHTS
- NIGHTSCAPING TWINKLELIGHTS
- UNDER WATER FLOODS
- S SWITCH LOCATION
- H HOUSE LIGHT LOCATION
- CEILING FIXTURE
- RECESSED CEILING FIXTURE
- RECESSED WALL WASHER FIXTURE
- WALL MOUNTED FIXTURE
- FLOURESCENT FIXTURE
- TRACK LIGHT FIXTURE
- S SINGLE POLE SWITCH +48" U.O.N.
- SD DIMMER SWITCH
- S3 S4 THREE WAY, FOUR WAY SWITCH
- COMBINATION LIGHT/EXHAUST FAN
- DUPLEX RECEPTACLE - ONE PLUG SWITCHED
- AREA USE LIGHT
- LOW VOLTAGE ELECTRICAL
- LINE VOLTAGE ELECTRICAL

LIGHTING ZONES

- A1.5 - SOUTH WEST WALL LIGHTS
- A1.4 - SOUTH EAST WALL LIGHTS
- A2 - LIBRARY COURTYARD ACCENT
- A3 - COURTYARD STRING LIGHTS
- B1.4 - TREE WALL LIGHTS & ACCENT EAST
- B1.5 - TREE WALL LIGHTS & ACCENT WEST
- C1.3 - SCULPTURE & ACCENT EAST
- C1.4 - SCULPTURE & ACCENT WEST
- D1 - DANA ST. WALL LIGHTS
- E1 - BACK WALL LIGHTS MIDDLE
- F1.3 - LIBRARY WALL LIGHTS & WALL
- F1.4 - MOON & ACCENT LIGHTS
- F2 - COURTYARD DOWNLIGHTS
- P1 - POND SUBMERSED LIGHTS
- B - MOON LIGHTS & ACCENT WEST
- C - WALL LIGHTS WEST
- P1 - POND LIGHT
- P2 - WATER FEATURE
- P3 - FILTER PUMP
- W - WALL LIGHTS
- X - ENTRY & ROSE GARDEN ACCENT WEST
- Y - ACCENT LIGHTS EAST
- Z - MOON LIGHTS EAST
- X-MAS 1 - RIGHT SIDE WALL
- X-MAS 2 - WEST SIDE TREE STARS
- X-MAS 3 - FRONT DOOR AREA
- X-MAS 4 FRONT EAST WALL
- X-MAS 5 - EAST SIDE TREE STARS

CONT.	ZONE	AREA CONTROLS	KIND	#of Fixtures	Watts Each	Fixture Load	Load + 20%	Max. Zone	Trans.
	B	Moon Lights Entry		12	35	420	504	504	1000
	Ca	Front Wall Lights West		7	35	245	294	396	600
	Cb	Front Wall Lights West		1	50	50	60	-	
	Cc	Front Wall Lights West		1	35	35	42	-	
	P	Pond Lights		LINE					LINE
	Wa	Garden Lights		5	35	175	210	558	1000
	Wb	Garden Lights		5		0	0	-	
	Wc	Garden Lights	F.	3	50	150	180	-	
	Wc	Garden Lights		4	35	140	168	-	
	Xa	Accent Lights (on zone C)		8	35	280	336	462	600
	Xb	Accent Lights (on zone C)	S.F.	3	35	105	126	-	-
	Za	Front Wall Lights East		3	35	105	126	336	600
	Zb	Front Wall Lights East		5	35	175	210	-	
	Zc	Front Wall Lights East		1	50	50	60	-	
	Ea	Column Lights		8	50	400	480	480	LINE
	Ya	Entry Lights	S.F.	3	35	105	126	354	600
	Yb	Entry Lights		2	35	70	84	-	
	Yc	Entry Lights		2	35	70	84	-	
	Yc	Entry Lights	F	1	50	50	60	-	
RA	H1	Front Door Lights		2	60	120	144	144	LINE
RA	H5	Front Side House Light		1	60	60	72	72	LINE

CONT.	ZONE	AREA CONTROLS	KIND	#of Fixtures	Watts Each	Fixt. Load	Load + 20%	Zone Load	Trans.
	D1	Dana St. Wall Lights		11	35	385	462	462	600
	E1	Back Wall Lights		10	35	350	420	420	600
	F1a	Moon Lights		5	35	175	210	520.8	600
	F1b	Moon Lights		2	50	100	120	-	-
	F1c	Wall Lights		3	35	105	126	-	-
	F1d	Courtyard Downlights		3	18	54	64.8	-	-
	A1a	Library Wall Lights		5	35	175	210	630	600
	A1b	Wall Lights		10	35	350	420		200
	A2	Wall Lights		2	35	70	84	84	200
RadioRA	A3	Tree String Lights		LINE					LINE
	B1a	Tree & Accent B		10	35	350	420	672	300
	B1b	Tree & Accent B	S.F.	6	35	210	252	-	-
	C1a	Sculpture & Accent B	S.F.	2	35	70	84	504	600
	C1b	Sculpture & Accent B	S.S.	10	35	350	420		200
	F2	Courtyard Downlights		6	35	210	252	252	600
	P1	Pond Submersed Lights		4	35	140	168	140	200
RadioRA	H2	Patio House Lights		4	40	160	192	192	LINE
RadioRA	H4	Library House Lights		2	40	80	96	96	LINE
RadioRA	H6	Garage House Lights		2	40	80	96	96	LINE
RadioRA	H11	Family room House Lights		2	40	80	96	96	LINE

LAGOS BLUE FOUNTAIN CAPPING 2" BULLNOSE
T.C. 100.15
CAPPING DETAIL
LAGOS BLUE WALL CAPPING 2" BULLNOSE
LAGOS BLUE RISER FACING
LAGOS BLUE BULLNOSE TREAD
LAGOS BLUE BULLNOSE TREAD
LAGOS BLUE RISER FACING
LAGOS BLUE WALL CAPPING 2" BULLNOSE
ENTRY LANDING SQ.FT. CALCS.
ENCLOSED LANDING AREA
ENTRY LANDING
STEP
RISERS
LAGOS BLUE DETAIL INSERT
LAGOS BLUE BORDER
FOUNTAIN DETAIL
SHADED AREA DELINEATES CAPPING FOR WALLS & FOUNTAIN

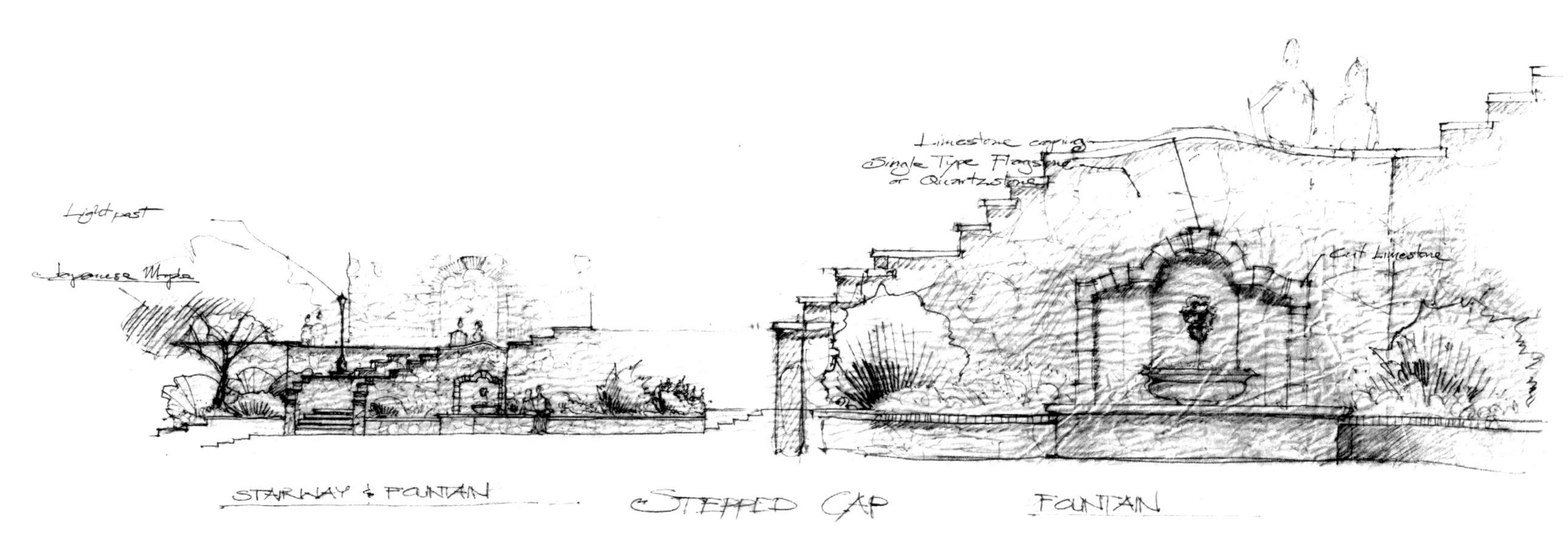
Light post
Japanese Maple
Limestone coping
Single Type Flagstone
or Quartzstone
Cut Limestone
STAIRWAY & FOUNTAIN
STEPPED CAP
FOUNTAIN

INDEX

Acres Wild Ltd

Acres Wild Ltd is an experienced, professional and widely respected garden design company run by Debbie Roberts BA (Hons) MSGD and Ian Smith BA (Hons) MSGD. After training in Landscape Architecture, Debbie and Ian chose to concentrate on the more intimate discipline of Garden Design, with its emphasis on close collaboration with individual clients. They established Acres Wild in 1988, and are both full members of the Society of Garden Designers. Their design philosophy is to respond sensitively to clients' needs and carefully integrate their house into the garden, and the garden into its surroundings, with plants and materials appropriate to the character and location of each particular site. This philosophy permeates all their design work, whether in town or country, at home or abroad.

STEVE TAYLOR

With over 18 years industry experience, Steve Taylor is one of Australia's leading landscape designers, specializing in high-end residential landscape design and construction throughout Melbourne, the Mornington Peninsula, regional victoria and interstate.

He founded COS Design in 2000 and since then, he and the team, have won more than 36 State design and construction awards along with the prestigious 2009 National residential landscape of the year.

The COS Team consists of Steve Taylor and 3 full time architects and their design studio is nestled in the leafy inner eastern suburbs of Melbourne Australia. Services offered include design consultations, design concepts, full architectural working drawings, 3D modeling and project management. COS Design also offer a full garden care and maintenance service.

VISTA Urban Planning and Landscape Design Office

VISTA Urban Planning and Landscape Design Office is established in 1985, by Dr. Turgay ATEŞ.

Since the beginning, the office completed many of the pioneering projects in Turkey like first pedestrianisation projects, first street landscaping projects and took part in many of the touristic resort projects in Turkey.

The Office is highly specialized on touristic hotel projects since the beginning. And it took part in some design projects in Germany, Russia, Iraq and Libya as well.

Design Focus Landscape Architecture & Construction

Design Focus is an International Landscape Architecture and Construction firm established in 1981 and located in Ben Lomond, California. They are currently involved in a wide range of projects including site planning, large scale commercial design projects, estate gardens, as well as historical renovation, habitat restoration, permaculture, and retreats. As award winning landscape architects they are recognized for their visionary aesthetic and ability to enrich and inspire the human spirit. They are LEED AP B C and D certified and experts at environmentally sensitive design and construction. Because they have installed many diverse large scale gardens consistently over the past 30 years, Design Focus has extensive knowledge and expertise in following the design process through to completion.

Lewis Aqüi Landscape + Architectural Design, LLC

Located in Miami, Florida Lewis Aqüí Landscape + Architectural Design, (LA2d) employs a working passion of varied, yet related disciplines to create outdoor living spaces of superb distinction. Their working knowledge of architecture, art, engineering, urban planning, sustainable design, green building, and horticulture mesh together to create visually pleasing and environmentally-sound solutions. They specialize in the creation and renovation of exquisite residential estates, boutique hotels, commercial buildings, public parks, and institutional site development projects throughout Florida and the Caribbean.

Kate Gould Gardens

KateGould
Gardens

Since 1998 Kate Gould has been working as a garden designer and she and her team have been creating award-winning gardens throughout London and South East England.

Commissioned by home-owners, architects, interior designers as well as property developers and commercial clients the team tackle all shapes and sizes of project from small city courtyards to large country estates; sleek and minimal or more traditional in feel and always with a mind to budget and expectations. Working closely with clients and contractors from plans to planting, often over the course of several seasons or even years ensures a garden design with stunning end results.

Kate is a regular exhibitor at the Chelsea Flower Show and has been awarded Gold medals as well the coveted 'Best in Category' twice, she also finds time to write regularly for *The Guardian*.

Landscape d.o.o.

Established by partners Gregor Vreš and Tina Demšar in 1998 the landscape architecture office has been cultivating a spirit of luxury while emphasising creativity, attention to detail and fully integrated landscape design. Through modern interpretation of heritage, traditional and autochthonous characteristics of space, the goal is to create simple and poetic landscapes that invigorate ways of living outside.

Landscape d.o.o. covers a broad field of work from landscape planning, nature preservation and restoration to designing outdoors of different scale – plazas, parks, public squares down to the smallest private gardens. Starting point of design are characteristics of space and client. In exploring proportions, light, materials, the focus is on unique, timeless designs that intertwine contrast, form and content.

DS Landscape

Being a pioneer in its field for 15 years, DS is a design office that has both theoretical and practical products on where architecture and landscape intersect. The main goal of the works is to create sustainable, practicable, well designed places with something to tell.

DS designs have the "setting of a microcosmos" as the main motivation. The design group perceives the complex fabric of today's multilayered, multicultural and interdisciplinary environment as a source and using this source, it sets up meaningful, clear, creative new places with the motto of combining high artistical skills with rational, economic and ethical values; both in national and international projects.

Considering nature as another source, the group creates parametrical designs through a syntactical perception of nature by filtering it.

DS provides design services ranging in scale from residential and urban projects, cooperating with distinguished architects, planners, engineers and especially ecologists.

The Friendly Plant (Pty) Ltd

Craig de Necker is an internationally-recognized, award-winning landscape designer and Managing Director of The Friendly Plant (Pty) Ltd.

Based in Johannesburg, South Africa, they design and install landscaped gardens, swimming pools, water features and designer outdoor spaces for clients based locally and internationally.

They have been featured in numerous books, newspapers, magazines and television broadcasts, including photographs of their work and interviews with their managing director.

Their clients include many high net worth individuals, the Big Brother reality television show and the current president of an African country.

They provide a complete turnkey service from creating the initial concept (2D CAD design and/or 3D design) through to the installation of the garden or outdoor space.

Merilen Mentaal

With fresh approach and quality know-how from the United Kingdom, Merilen has graduated from the Oxford College of Garden Design, one of the top garden designs diploma courses in the world.

Appreciating modern materials, she tends to prefer simple, geometric, contemporary functionalism, with clean and straight lines, creating a balanced special flow within exciting outdoor rooms.

Combined previous experience in business, management, event marketing and corporate travel, gives an advantage to understand different needs, putting them in right perspective financially as well as functionally. Staying highly creative, absorbing the influences of different cultures, she believes in constant improvement, managing projects on an international level. She works in Estonia, United Kingdom and Switzerland but is open for new challenges in other countries.

Rolling Stone Landscapes

Dean Herald is one of Australia's leading landscape designers. Principal Designer and Managing Director of Rolling Stone Landscapes, he started the company at age 19. Since then he has achieved the pinnacle of the landscape design industry winning a gold medal at the prestigious Royal Horticultural Society Chelsea Flower Show in London along with being named New South Wales and Australian Landscaper of the year. Dean has built a reputation in the prestige residential market designing resort-style external spaces and taking the concept of outdoor living and entertaining to a dynamic new phase. Dean and his team have built seven show gardens being awarded a gold medal on each occasion, along with three design excellence awards and the Best in Show award at the Melbourne International Flower & Garden Show.

Dean's work has been featured in leading Australian television programs and is showcased in his books "Resort Style Living" and "21st Century Residential Landscape Design".

Cherry Mills

Cherry's practice near Guildford, is broad based and caters for family gardens through to large country estates, in traditional or contemporary style. She works mainly in Surrey but can also undertake commissions abroad. She trained in garden design at Merrist Wood College, Guildford, qualifying with distinction in 1994. She succeeded in becoming a Registered Member of The Society of Garden Designers in 2000 and is featured on the Society's list of approved designers.

Her aim is to create beautiful and practical gardens which suit her clients' taste and lifestyle. She hopes that she can pass on to her clients her love of plants, with their infinite range of shape, color and texture.

Eckersley Garden Architecture

eckersley
Garden Architecture

Eckersley Garden Architecture is a boutique landscape design firm based in Melbourne that brings a revitalised approach to landscape creation from the renowned stable of Rick Eckersley. Long time associates and recent partners Scott Leung and Myles Broad along with Kathryn Green join Rick to bring knowledge, innovation and passion to a changing industry.
Eckersley Garden Architecture receives commissions Australia wide and internationally. Their client base is as varied as their garden designs, ranging from small residential, to commercial multi residential, to country retreats. They also work closely with leading Australian architects.
The business operates from two locations - the principle office in the Melbourne inner city suburb of Richmond, and a secondary office in Flinders, a rural coastal area of the Mornington Peninsula.

Terragram Pty Ltd

Whilst the initial impetus for Terragram's existence came from winning a competition (1985), another incentive was to have a working platform that would engage in a critique of the then prevailing pragmatic approach to designed landscape, pursued by most practices in Australia in the mid eighties.
Despite an undeniably increasing scope of tasks solved by landscape architects today, they still consider the narrow definition of landscape architecture quite limiting, especially if forced to operate in pre-determined situations with a scope further reduced by political expediency and their value system. The profession appears hopelessly suspended between sustainability and impotence and cultural sustainability does not seem to be ever entering the formulaic approaches.
They do not consider themselves as a traditional office. The French word 'atelier' better describes the atmosphere and their inclination to experiment, cherish intuitive responses, invite the unexpected and scary, and to technologically innovate. Despite the modest size of the company, their interest range also includes stage design, sculpture, graphics, furniture and educational activities. Their cultural curiosity led them to work in different countries outside Australia.

Ecocentrix

Ecocentrix was founded on the fundamental premise that - the quality of the experience and function of landscapes is achieved by understanding inherently "what is" and "what is wanting", and that quality of life is a reflection of the quality of the landscape.
Their body of work exemplifies great stylistic range and restraint produced with consistently high quality. Their projects are immediately mood altering, celebrating the sensual and tactile temperament that is the fabric of landscape.
Their design creates the ground for celebrating the cycles of all life, and is the foundation of regional identities enveloping cultural distinctions. It reinforces what is powerful and enhances what is weak. Ecocentrix endeavors to "Enrich Life Through Design".

IAN KITSON

Ian Kitson has been designing gardens and landscapes for over 30 years. His work has a precise but free form character which provides a bespoke and individual identity to the gardens and landscapes he designs. Ian has qualifications in landscape architecture, architecture and the conservation of historic garden and designs for a wide variety of private, public and commercial clients both in the UK and internationally. His work has won several awards and has been featured in books and on TV. His practice is based in central London.

Three Sixty Design

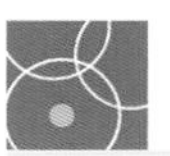
three sixty desig

They believe in dynamism. They believe in inspiration. They believe in collaboration. They believe in flexibility. They are a landscape architecture studio and they believe their work transcends the ground plane.
They work engages, drives and motivates. They create dynamic and innovative spaces that truly fit each and every client; perfectly bespoke solutions.
They derive inspiration from everywhere. They consume, obsess and collect; all to deliver a truly different and specific spatial experience. They visit inspiring places, collect imagery and experiment daily to truly keep the creativity flowing.
Their work is a master study in collaboration. They work seamlessly with their architects, designers, engineers, consultants, and more importantly, their clients. They don't assume or impose; they reflect and examine. They are designers with technical expertise, not dictators of taste.

Alex Hanazaki

Architect by training and landscape designer by vocation, Alex Hanazaki is above all an esthete.
The childhood and youth in Presidente Prudente gave the roots needed to build projects with the foot on earth, impregnated with emotional references, but it's his cosmopolitan soul that turns into amazing scenarios. Alex extracts of nature shapes, colors and textures with his hands, translating them to a modern landscape. Private gardens of pure luxury, minimalist environments in concept, care, made of details that flirt with purism and their meanings. Owner of an Office in Sao Paulo, in tune with his perfectionism, Hanazaki is considered one of the great names of modern landscape in Brazil. To him, landscaping is also quality of life. But beyond that, the challenge is to create for each project gardens as works of art.

JOHN FELDMAN

A native of Los Angeles, John Feldman studied at the College of Architecture and Environmental Design at California Polytechnic State University, San Luis Obispo, where he received his Bachelor of Science degree in Landscape Architecture. Before beginning his professional career he embarked on an extensive overseas study program traveling throughout China and Southeast Asia. Through both his own individual study and his collaboration with professional offices and universities, he sought to explore, in depth, the social and cultural impacts on architecture, urban planning and the natural environment.
Feldman has been involved in a wide range of project types, including commercial retail, street improvements, museum and institutional, public open spaces, multi-family housing/mixed use planning, residential gardens and estate master planning. His skills reflect the diversity of the projects and his ability to provide expertise in design, public relations, technical problem solving, scheduling and budgeting issues.
Feldman enjoyed his tenure while at some of the most prolific design firms in Los Angeles. As Director of the Landscape Studio at KAA Design Group, he directed all aspects of design, management, and construction administration for the range of opportunities at the firm. Feldman credits having honed his skills in design and theories in "regional contextualism" while at the firm Nancy Goslee Power and Associates. In addition to strong business skills developed while having formerly operated the firm Garness / Feldman - Architecture + Landscape, Feldman brings extensive international design experience from having completed landscape projects in many countries around the world.
Licensed in the States of California and Hawaii, Feldman enthusiastically leads Ecocentrix, Inc. with vigor, vision, and evolved paradigms, with resulting design investigations spanning traditional to progressive - where ever the firm's work takes them.

laND30

laND30 is a young and energetic studio with a multidisciplinary team of professionals sharing artistic and environmental sensibilities as well as a passion for design. It has its epicenter in Spain and spreads through Europe with its collaborators.
The approach is holistic and with attention to detail. Starting with a solid concept and developing it like a sculptor that chisels through a rough stone till a polished final proposal is achieved. Studying the context of the project is important: climate, ecology, culture, history, etc. This leads to a careful selection of soft and hard materials, shapes, colours and aesthetics to fit into the context of a project.
laND30 has worked at various landscape scales, from the small garden to larger scale urban projects. Having produced a few notable private gardens in collaboration with A-cero architects.

Lutsko Associates

LUTSKO ASSOCIATES
landscape

Founded in 1981 by Ron Lutsko, Jr., Lutsko Associates works on award-winning projects throughout the world in all areas of professional practice: residential, commercial, civic and institutional. Their clients include public agencies, non-profit organizations, educational institutions, citizens groups, businesses and individuals.
At Lutsko Associates, They are dedicated to high quality, forward thinking landscape design and are known for their effectiveness in meeting the needs and desires of a unique and varied clientele. They believe in a multi-disciplined approach to design, typically collaborating with architects, planners, artists, scientists, and engineers. This collaborative process brings greater social meaning, environmental sensitivity and beauty to their landscapes.

Marpa Landscape Design Studio

Marpa Landscape Design Studio was founded in 1974 to bring peace to the world one garden at a time. It was founded by Martin Mosko, a Zen Buddhist monk. Over the years, Marpa has won 52 awards for excellence in design, and been published in books and magazines throughout the world. They are currently building a Hindu monastery in Kauai, Hawaii, and a temple in south India. :"The Garden of Infinite Compassion" in Sun Valley, Idaho was selected by the US. Botanic garden, as one of the top spiritual gardens in the US.

John Nash

John Nash is a specialist Garden Design and Landscaper, a fully qualified member of the SGD, The Society of Garden Designers and a BALI, the British Association of Landscape Industries.
His design philosophy respects the environment and his style leans towards contemporary but his versatility enables him to be equally comfortable with traditional styles often drawing on classic elements. He is adept at finding solutions for spaces so that there is harmony between hard landscaping and plants. He is equally comfortable with small town Courtyards to large country gardens and influences derive from the Mediterranean, Japan, and Islam.

Hendy Curzon Gardens Ltd.

Hendy Curzon Gardens Ltd. creates modern and timeless outdoor spaces and places that directly reflect individual clients styles and fulfill their needs. Every garden created is considered as a true extension of the home. As a diverse and dedicated team of designers and landscapers, Hendy Curzon Gardens Ltd prides themselves on a personal and long-term commitment to the landscapes and gardens that they create. Each exterior space designed and constructed is unique and a direct result of the client brief. Company established in 2004.

Jim Fogarty Design Pty Ltd

Jim Fogarty graduated from Melbourne's Burnley Horticultural College in 1992 with a passion for designing gardens. Today he designs gardens for private and commercial clients as well as designing Show Gardens at world famous garden shows.
Jim has won numerous awards in Australia and overseas, including 11 Gold Medals.
In May 2009, Jim won the award for Best Design for his Australian Show Garden at the World Garden Competition in Shizuoka Prefecture, Japan. At the Chelsea Flower Show in 2011, Jim designed an Australian Show Garden for the Royal Botanic Gardens Melbourne that won a gold medal on the Main Avenue. At the 2011 Gardening World Cup, Japan, Jim was awarded 'Best in Show' winning the Gardening World Cup for Australia.

ASPECT Studios

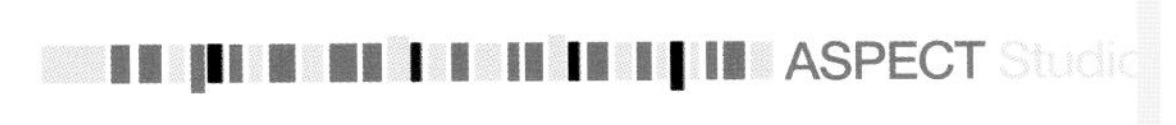

ASPECT Studios is a group of design studios united through a philosophy that delivers innovative landscape architecture, urban design and digital technologies.
Since it's beginning, ASPECT Studios has grown to over 55 people throughout their Australian and China studios. They have established a reputation for design-led solutions and are recognised as a company with the capability to deliver world leading design excellence through creative and sustainable projects.

Hothouse Design

Hothouse Design is the branch office of Italy MEDITERRANEO DESIGN, specialized in garden design, landscaping and related services. In HOTHOUSE DESIGN, a professional designer team, mainly composed of senior designers from Italy Headquarter, as well excellent domestic, led by Chief Designer, Mr. UGO WANG, has completed a number of excellent garden and landscaping projects in Shanghai and other cities in China. Now HOTHOUSE DESIGN is highly recognized by many real estate developers and owners of private villas. The wide variety enables this team adapt to increasingly demanding requirements of clients.

Scott Brown Landscape Design

Formed in 1993, this garden design practice focuses on high-end private, residential landscape design. The scope of the projects undertaken can encompass the design and project management of all aspects of the outdoor environment including swimming pools, outdoor entertaining areas and pavilions, water features, and lighting as well as the gardens themselves. Projects undertaken include every size and setting, ranging from high-rise inner city balconies, converted warehouse landings, inner-city courtyards, suburban gardens and outdoor spaces (large and small), through to large-scale rural properties.

The practice's principal, Scott Brown, has accumulated an impressive collection of both industry and public awards. Following University (Melb.) where Scott was awarded a first-class honours degree in Earth (and Horticultural) Sciences, Scott completed his adult apprenticeship in Landscape design and construction, during which he was awarded "Victorian Landscape apprentice of the year" in every year of his apprenticeship. This was then followed up with graduate diplomas/certificates in Landscape Design, Urban Planning, and Hydrology.

Scott Brown Landscape Design has been awarded 'highly commended' by the Australian Achiever Awards for Customer Satisfaction.

Studio H Landscape Architecture

studio h
landscape architecture

Studio H Landscape Architecture brings its passion for creative design solutions, innovative design details and quality execution to make sure each of its projects exceeds the client's expectations. Its work is recognized for marrying its creative aesthetic design with the latest sustainable products and practices all while staying on budget and ensuring the client receives the best possible product for their dollar.

ACKNOWLEDGEMENTS

We would like to thank everyone involved in the production of this book, especially all the artists, designers, architects and photographers for their kind permission to publish their works. We are also very grateful to many other people whose names do not appear on the credits but who provided assistance and support. We highly appreciate the contribution of images, ideas, and concepts and thank them for allowing their creativity to be shared with readers around the world.